MATHEMATICS EDUCATION IN CHINA:

ITS GROWTH AND DEVELOPMENT

The MIT Press

Cambridge, Massachusetts, and London, England

MATHEMATICS EDUCATION IN CHINA:

ITS GROWTH AND DEVELOPMENT

Frank Swetz

Library of Congress Cataloging in Publication Data
Swetz, Frank.
    Mathematics education in China.

    Bibliography: p.
    1. Mathematics--Study and teaching--China.
I. Title
QA14.C6S95   510'.7'1051   73-18274
ISBN 0-262-19121-0

學習數學與天文學是今日學者當務之急

" . . . it is incumbent on our scholars to learn the rudiments of mathematics and astronomy." Tz'u-hsi, Empress Dowager of China, March 5, 1867. The royal command that initiated modern mathematics education in China.

Chinese Monetary Units and Measures

Referred to in Text

Currency

    tael = 37.8 gm of silver = U.S. $1.68
    candareen = 5.833 gm of silver = U.S. $0.26
    mace = 3.78 gm of silver = U.S. $0.17
    Mexican dollar = 0.72 tael = U.S. $1.04
    yuan = U.S. $0.40
    fen = .01 yuan = U.S. $0.004

Weight

    catty = 0.605 kg = 1.333 lbs.
    picul = tan = 60.48 kg = 133.3 lbs.

Capacity

    tou = 10.3 liters = 0.293 bushel

Land Measures

    fan = 0.14 inches
    ch'ih (Chinese foot) = 14.1 inches
    pu   = 70.5 inches
    li (Chinese mile)   = 0.40 std. mile
    mou (Chinese acre)   = .165 acres

ACKNOWLEDGMENTS

The task of documenting and assessing the trends
in mathematics education in the People's Republic of
China is extremely difficult. Political, cultural and
linguistic barriers stand in the way of a Western
researcher attempting such a project. By necessity, the
researcher must seek help from many sources. This was
the situation in which I found myself when I began this
study. Fortunately, my quest for information was marked
by cooperation and kindness. I would like to take this
opportunity to express my thanks to some of the people
who assisted and encouraged me through the completion of
this work.

I am indebted to the following professors at
Teachers College, Columbia University: Bruce Vogeli, for
nurturing my interest in international mathematics education; Chang-tu Hu, for exposing me to the study of Chinese
education; and the late Myron Rosskopf, for his many useful editorial comments. My thanks to the persons who
supplied me with research information, particularly:
Professor Steward Fraser and John Hawkins, George Peabody
College for Teachers; Robert Barendsen, United States
Department of Health, Education and Welfare; Professor
Richard Sorich, East Asian Library, Columbia University;

Professor Peter Seybolt, University of Vermont; and Doctor R. C. Price and Professor Chu-sam Tsang. Calligraphic rendering of the Tz'u-hsi quote was by Shih-chuan Chen, Professor of Humanities and Oriental Civilization, Capitol Campus, Pennsylvania State University. Assistance in translating Chinese materials was provided by Messrs. T. I. Kau, Peter Au and Hsi-sheng Nieh.

Special thanks to the library staffs at Capitol Campus, Pennsylvania State University and the Orientalia Division of the Library of Congress for their assistance in securing reference material. Finally, I wish to thank the East Asian Institute, Columbia University, for providing me with a grant that made much of my research possible.

Capitol Campus   Frank Swetz
Pennsylvania State University
Middletown, Pennsylvania
July, 1973

TABLE OF CONTENTS

Chapter

Chapter

Chapter

LIST OF TABLES

PUBLISHER'S NOTE

This format is intended to reduce the cost of publishing
certain works in book form and to shorten the gap between
editorial preparation and final publication. The time
and expense of detailed editing and composition in print
have been avoided by photographing the text of this book
directly from the author's typescript.

The MIT Press

MATHEMATICS EDUCATION IN CHINA:

ITS GROWTH AND DEVELOPMENT

INTRODUCTION

Since the middle of the nineteenth century, China
has been engaged in a continuous struggle to free herself
from the intellectual and economic bondage imposed by a
traditional heritage. The Chinese are attempting to
replace their agricultural subservience with a technological self-sufficiency. The primary instrument of this
transformation has been education, particularly scientific-
based education. One of the Manchu government's responses
to the West's initial impact upon China was to demand
that its scholars become proficient in the languages and
sciences of the "barbarians." By the first decade of the
twentieth century, this fledgling effort at modern education had developed into a national school system. Educational activity increased during the following Republican
Period with the focus of education becoming more nationalistic. The study of science and mathematics was
intensified to increase the rapidity of the "Middle
Kingdom's" advance into the industrialized world family
of nations. This academic emphasis was continued through
the Kuomintang Era and into the present Communist Period
of Chinese history.

Perhaps no modern Chinese government has been more
dedicated or effective in developing China's scientific

potential than the Communists. The importance and objectives of education in the People's Republic of China were clearly defined in the new state's Common Program of October, 1949:

> Article 41: The culture and education of the People's Republic of China are new democratic, that is, national, scientific and popular. The main tasks for raising the cultural level of the people are: training of personnel for national construction work; liquidating of feudal, comprador, Fascist ideology; and developing of the ideology of serving the people.
>
> Article 42. Love for the fatherland and the people, love of science, and the taking care of public property shall be promoted as the public spirit of all nationals of the People's Republic of China.
>
> Article 46. The method of education of the People's Government shall reform the old educational system, subject matter, and teaching methods systematically according to plan.
>
> Article 47. In order to meet the widespread needs of revolutionary work and national construction work, universal education shall be carried out. Middle and higher education shall be strengthened; technical education shall be stressed; the education of workers during their spare time and the education of cadres who are at their posts shall be strengthened; and revolutionary political education shall be accorded to young intellectuals and old-style intellectuals in a planned systematic manner.[1]

The guidelines for education are declared to be nationalism, popularism and scientism.

---

[1]Stewart Fraser, Chinese Communist Education: Records of the First Decade (Nashville: Vanderbilt University Press, 1963), pp. 83-84.

In their drive to foster a scientific spirit in the New China, educators radically increased the academic importance of the studies of foreign languages, physics, chemistry and mathematics. Of all the science-oriented subjects in the Chinese curricula, mathematics was deemed the most important. During the Second National People's Congress held in Peking in 1960, Lu Ting-yi, Vice Premier of the State Council, speaking on projected education reforms, was explicit in his educational priorities. For Lu, there were only two basic academic disciplines of importance:

> If language and mathematics are properly mastered, it becomes relatively easy to master physics, chemistry, biology, history, and geography. The joint efforts of all teachers are needed in enabling the students to master languages and mathematics properly . . .[2]

This theme was again reinforced in a 1962 Hupeh Jih-pao (Hopeh Daily) editorial in which the purpose of secondary education was defined as that of training "a reserve labor force" together with preparing students for higher studies.[3] According to this editorial, the basic subjects to be studied were national language, mathematics

---

[2] Current Background, No. 630 (Hong Kong: United States Consulate General, August 8, 1960); Translated from People's Daily, June 15, 1960.

[3] Surveys of China Mainland Press, No. 2669 (Hong Kong: United States Consulate General, 1962).

and a foreign language. Thus the acquisition of a
mathematical proficiency amongst the youth of China,
while always considered in high regard by modern Chinese
governments, has now been elevated to the status of a
national priority!

While several works have investigated education
in Communist China in toto,[4] few researchers have
endeavored to examine the academic potentialities of the
specific disciplines taught in Chinese schools. Those
who have, such as Morrison or Doolin and Ridley,[5] have
limited their scope of inquiry to subjects that readily
lend themselves to the promotion of nationalistic or
political designs, i.e., history and literature. These
investigations have provided some understanding of the
ideological foundations of the People's Republic of
China and its success in promoting the first two aims
of its educational credo--nationalism and popularism.
The third aim as stated in the Common Program, scientism,

---

[4]Fraser, op. cit.; Chang-tu Hu, Chinese Education
Under Communism (New York: Teacher's College Press,
Columbia University, 1962); R. F. Price, Education in
Communist China (New York: Praeger Publishers, 1970).

[5]Ester Morrison, "A Comparison of Kuomintang and
Communist Modern History Textbooks," Papers on China,
Harvard Seminars, Vol. VI (Cambridge, Mass., March,
1952), pp. 3-45; Dennis Doolin and Charles Ridley, The
Genesis of a Model Citizen in Communist China (Washington, D. C.: United States Office of Health, Education
and Welfare, June, 1968).

and its effect on the schooling of China's young has received little attention in research literature. Such an omission permits the question of China's scientific and industrial potential to remain moot. By the Chinese government's decision to base the scientific studies of its young upon a strong mathematics background, an avenue of evaluation has been provided to examine this question. The success of the People's Republic of China in instituting rigorous and widespread mathematics education reforms will directly reflect on its future scientific output. The question of Chinese mathematical reforms, their content and acceptance, has significance in determining the People's Republic of China's possible position as a world power.

The main purpose of this study is to attempt to document and assess the trends in mathematical education at the primary and middle school levels in China. In the process of evaluating the effects of Chinese mathematical reforms, comparisons will have to be made. China's present state of economic development combined with her particular cultural heritage make comparison with Western educational norms almost meaningless. The effectiveness of mathematics reforms lies in their ability to eliminate previously existing deficiencies and satisfy the contemporary needs of Chinese society. Thus in seeking to understand the objectives, techniques and content of mathematics in Chinese schools, I am compelled to investigate the conditions that forged them. This study will then concern itself with the evolution of mathematics education in the People's Republic of China as resulting from a series of social and political pressures.

Four thousand years of Chinese history has produced prejudices to intellectual and social reforms that all modern governments of China had to contend with. Attempts to overcome the bias of a traditional heritage are very much in evidence in the policies of the People's Republic. This phenomenon has nicely been summed up by John Fairbank:

> China's long history has him [Mao] in quick-
> sand--as he struggles, he becomes more immersed
> in the attitudes and dreams inherited from
> China's past.[6]

In order to fully appreciate mathematics reforms that may appear alien to Western experience, a substantial historical perspective will be established. This perspective will trace the evolution of Chinese mathematics education through the modern period, i.e., from 1860 to the present. Educational progress will be restricted to a consideration of government primary and middle schools

---

[6]John Fairbank, China, The People's Middle Kingdom, and the U.S.A. (Cambridge, Mass.: Belknap Press of Harvard University, 1967), p. 4.

and selected missionary institutions. For the period after 1949, the scope of the study will be broadened to include the Communists' wider spectrum of proletarian educational institutions. Tracing of this evolution will then culminate in a consideration of the specific reforms made under the present government of China. In particular, answers will be sought to the following questions:

1. How pervasive was Soviet influence in the upgrading of Chinese school mathematics?

2. Were "special schools" for mathematically talented students established in the People's Republic of China following Soviet models?

3. Why did the attempts to institute "Olympiad"[7] type mathematical examinations in China falter?

4. What was the modern mathematics program suggested for national adoption by the Peking Normal University in 1960?

5. What are the present effects of "The Great Cultural Revolution" on mathematics education in China?

6. How are "practical applications" being introduced into mathematics teaching?

7. Are the Chinese effectively solving their teacher shortage?

8. To what extent are the psychological processes of mathematics learning being considered in mathematics instruction?

Secondary tasks of this study will be: a consideration of the effects of political indoctrination on

---

[7]National government sponsored examinations used to detect mathematical talent in the Soviet Union.

mathematics instruction and the relevance of Chinese
mathematics teaching practices to other developing
nations. Every aspect of education in Mainland China,
while imparting academic knowledge, also functions as an
avenue of political indoctrination. Indoctrination supplementing mathematics instruction may obscure the mathematical principles being taught and provide a poor learning situation. This study will therefore also attempt to
determine the extent to which mathematics teaching has
been used for proletariat indoctrination.

Social and economic conditions in the People's
Republic of China in many respects parallel those of
other developing nations. Successes or failures of various educational practices in mathematics education might
have implications for educational reform in other countries of the Third World. These practices will be
isolated and commented upon.

The political and social isolation of Mainland
China presents a severe obstacle to the Western researcher.
In the absence of personal observation of the Chinese
mathematics education scene, the nucleus of material
gathered in this study will be distilled from a variety
of sources: official Chinese educational documents,
textbooks and journals; previous research done on various aspects of Chinese society and institutions, including
education; and interviews either in person or by

correspondence with people directly involved with Chinese mathematics education.

Primary Chinese sources of information will be obtained from the following publications:

Surveys of China Mainland Press (S.C.M.P.), Hong Kong

Current Background, Hong Kong

China News Analysis, Hong Kong

Shu-hsüeh T'ung-pao[8] (Mathematics Bulletin), Peking

Chung-hsüeh Shu-hsüeh (Middle School Mathematics), Changsha

Hsin-li Hsüeh-pao (Acta Psychological Sinica), Peking

In the analysis of the material available from these sources, care must be taken to avoid permitting the tenor of the political rhetoric from obscuring the facts sought. Highly charged political indoctrination grates against the ear of the uninitiated researcher. Prolonged exposure to such material has helped overcome this barrier.

The tasks undertaken in this study are not easy, but their importance and implications would seem to warrant the effort involved. In the procedures suggested for this research, no one alone is going to provide

---

[8]The transcription of Chinese words in the text will generally follow the Wade-Giles system of pronunciation. Exceptions to this policy will include names of official publications cited whose titles have been romanized by the Chinese using their Han-yu Pin-yin system. Thus Mathematics Bulletin in Wade-Giles will be Shu-hsüeh T'ung-pao but in Han-yu Pin-yin, Shuxue Tongbao.

substantial information on which to draw conclusions or, more importantly, infer predictions. Taken together, however, in the proper historical perspective, a mosaic-like image should appear. It is from this resulting image that the conclusions concerning mathematics education in the People's Republic of China will be drawn.

CHAPTER I

HISTORICAL PERSPECTIVES: THE FORMATIVE STAGE OF
MATHEMATICS EDUCATION IN CHINA

Development of Chinese Mathematical Thought

The origins of Chinese mathematics are obscured by
legend and mysticism. A variety of mythological beings
and creatures are credited with assisting in the development of the mathematical sciences in China. Legendary
emperor Huang-ti (the Yellow Emperor, twenty-seventh
century B.C.) emerges as the first patron of mathematics.
His ministers, Tai-mao and Li-shou, devised a sexagesimal
scheme of numeration and formulated an "art of numbers"
or arithmetic. Royal predecessors, Fu-hsi and Yü,
founded systems of numerical and geometrical computation.
Fu-hsi's theories were inspired by diagrams, Pa-kua,
provided by a dragon-horse.[1] Yü, considered the patron
saint of hydraulic engineers, experienced a similar
phenomenon when a divine tortoise emerged from the Yellow
River displaying a numerical diagram, Lo-shu, on its
back.[2] These diagrams occupied a prominent position in

---

[1]David E. Smith, History of Mathematics (New York:
Dover Publications, reprint of 1923 edition), vol. I, p.
26.

[2]Yoshio Mikami, The Development of Mathematics in
China and Japan (New York: Chelsea Publishing Co.,
reprint of 1913 edition), p. 3.

early Chinese divination and sooth-saying, and were later
to be known as "Magic Squares."[3]

From earliest times, the philosophical and metaphysical properties of numbers were employed by the
Chinese in their attempts to achieve spiritual harmony
with the cosmos. Scholars, impressed by the orderings of
nature, which to them seemed to obey some mysterious and
unchanging laws, attempted to incorporate similar orderings in their theories of human conduct and existence.
They felt that man's undiscordant existence depended upon
numerically specified actions and obligations. Thus,
traditional writings are explicit when they categorize and
enumerate three forms of obedience or the six virtues.

Historically, the earliest concrete evidence of
mathematical activity in China is given by the inscription
on a set of oracle-bones (fourteenth to eleventh centuries
B.C.), the symbols of which indicate a system of numeration.[4] This system included tally and code symbols,
utilized positional value, and was decimal in conception.
Later systems were to be more graphic in nature to accommodate computing schemes employing a set of rods. At

---

[3]For further information on magic squares see
Schuyler Cammann, "The Evolution of Magic Squares in
China," American Oriental Society Journal (1960), 80:
116-24.

[4]Joseph Needham, Science and Civilization in China
(London: Cambridge University Press, 1959), vol. 3, p.
5.

various times in Chinese history, several different
systems of numeration were in existence simultaneously,
each corresponding to a different social application.

The oldest extant Chinese mathematical text known
is the Chou-pei Suan-ching, dated at approximately 300
B.C. Although mainly geometrical in content, it provides
evidence that computers of the time could perform operations on fractions in accordance with modern practices.
There is also evidence that processes for root extraction
were known. The Chou-pei's most interesting contribution
is an exposition on the properties of the right triangle,
including a demonstrative proof of the Pythagorean Theorem.
Slightly later in date of origin, but with far greater
influence for the propagation of Chinese mathematics, is
the Chiu-chang Suan-shu or Nine Chapters on the Mathemat-
tical Art. This thesis appears to be a compilation of the
mathematical knowledge known in China up to the third century A.D. Reflected in the titles of its nine sections is
the utilitarian nature of the mathematics of the time:

1. Surveying of Land
2. Millet and Rice
3. Distribution by Progressions
4. Diminishing Breadth (Area Problems)
5. Consultations on Engineering Works
6. Impartial Taxation
7. Excess and Deficiency (Rule of False Position)
8. The Method of Calculating by Tabulation
9. Right Angles.[5]

---

[5]Ibid.

It is a manual explaining how to obtain solutions for the
particular mathematical problems presented. Little indication is provided as to the manner in which the rules
for solution were obtained. The context of the solutions
is algebraic-arithmetic and reveals an empirical methodology. Chinese mathematicians lacked a system of algebraic
notation; consequently, the problems and solutions considered are presented in a literary form. Whatever the
Chiu-chang's limitations, this manual indicates a high
degree of computational efficiency, including a rule of
false position, a root extraction process similar to
"Horner's Method,"[6] a facility in obtaining solutions to
systems of simultaneous equations, and a use of negative
numbers. These two works exemplify the trends of Chinese
mathematical thinking for the next two millennia. Predominant in this thought was:

1. Mathematics as a body of inductively conceived knowledge resulting from concrete
experimentation, rather than the application of a deductive scientific method.

2. Mathematics as a collection of computational
techniques formulated through an extensive
reliance on mechanical computing aids and
designed to solve concrete problems, usually
problems arising from the necessary functioning of the state.

---

[6]Wang Ling and Joseph Needham, "Horner's Method in
Chinese Mathematics: Its Origins in the Root Extraction
Procedures of the Han Dynasty," T'oung Pao, Leiden (1955),
43: 345-388.

3. A geometry founded in the mensuration needs
    of an agrarian society and never elevated
    to a systemized study as in the West.

Native Chinese mathematics attempted to resist, at least in principle, the transgression of foreign mathematical concepts that began to be transmitted from India in the third century. This opposition by Chinese scholars was based in part on the notion that mathematics could not be viewed as an abstract intellectual activity transcending reality. Shen Tso-che, a scholar of the twelfth century, expresses his opinion on this matter:

> That which is vague and obscure can have no place
> in matters connected with number and measure.
> Whether the numbers or dimensions be large or
> small, problems can all be solved and the answers
> distinctly stated. It is only things which are
> beyond shape and number which cannot be investi-
> gated. How can there be mathematics beyond the
> reach of shape and number?[7]

Official opposition to foreign mathematical theories ended in the fourteenth century when the establishment of a Muslim Astronomical Bureau in Peking was sanctioned by the throne. Despite this isolation, Needham speculates that an Arabic translation of Euclid's _Elements of Geometry_ existed in China by the end of the thirteenth century.[8] If this was the case, it appears there was no perceptible effect on the geometrical thinking of Chinese

---

[7] Needham, _op. cit._, p. 88.

[8] _Ibid._, p. 105.

mathematicians in the following centuries.

The greatest advances of indigenous Chinese mathematics occurred during three periods: the Han, the late Sung, and the early Yuan. The Chou-pei and Chiu-chang, with their wealth of utilitarian mathematical knowledge, were both products of the Han period. Important Sung and Yuan writings included Ts'e-yüan Hai-ching (Sea Mirror of Circle Measurements, + 1248); Hsiang-chieh Chiu-chang Suan-fa Tsuan-lei (Analysis of the Rules in the "Nine Chapters," + 1261); Suan-hsüeh Ch'i-mêng (Introduction to Mathematical Studies, + 1299); and Ssŭ-yüan Yü-chien (Precious Mirror of the Four Elements, + 1303). Although some of the theses of the Sung and Yuan eras concern geometry, the major contributions to mathematical thought were algebraic. Indeterminate analysis, treatment of equations of higher degrees, and refinement of previously known algebraic techniques highlighted the achievements of these periods. A diagram in the Ssŭ-yüan indicates that the "Pascal Triangle" method of determining binominal coefficients was known in China at this time.[9] Even though advances were made in computational techniques, the transcription of this knowledge was still hindered by a lack of adequate symbolism.

---

[9]Carl B. Boyer, A History of Mathematics (New York: John Wiley and Sons, 1968), p. 228.

The historical period of indigenous mathematical accomplishment in China came to an end with the arrival of Jesuit missionaries at Peking in 1601. By this time, native mathematics had lapsed into a state of dormancy and decay. Accepted as a classical scholar, the Jesuit Matteo Ricci, a former pupil of Christopher Clavius, translated two of Clavius' mathematical works into Chinese. Euclides Elementorum Libri XV became the Chi-ho Yüan-pên, but included only the first six books of Euclid, and Epitome Arithmetica Practicae was translated into the T'ung-wên Suan-chih. Jesuit predecessors of Ricci introduced the principles of logarithms to the Chinese and assisted in refining native theories of trigonometry. This transfusion of European knowledge revitalized the ailing Chinese sciences. The resulting period of mathematical productivity saw the publication of Chih-shui I-chên (Pearls Recovered from the Red River), a text reflecting Chinese ethnocentrism by its claims of Chinese origin of foreign accomplishments.

Through its various periods of development, Chinese mathematical thought retained the characteristics exhibited in its initial theses: empiricism and utilitarianism.

The Position of Mathematics in the
Traditional Educational System

Educational institutions are first mentioned in
the history of China during the Shang dynasty (ca. 1523-
ca. 1028 B.C.). During this period, the legendary
emperors Shun and Yao were credited with establishing two
schools near the imperial palace: The Shang-hsiang for
the education of princes and the Hsia-hsiang for the education of the common people.[10] The curriculum devised by
emperor Shun had three divisions: the five human relationships, the three religious ceremonies, and music.
Such an education enabled its recipient to live in harmony
with himself, his fellow men, and with the spirits.
Apparently, at this time, no consideration was given to
formal literary or scientific studies. In order to assess
the results of this education and insure that his officials were indeed harmonious beings, Shun instituted a
civil service examination.

> Emperor Shun every three years examined the merits
> of the officials. After three examinations, he
> promoted the intelligent and degraded the incom-
> petent. Thus all affairs flourished.[11]

---

[10]Ping Wen Kuo, The Chinese System of Public Education (New York: Teachers College Press, Columbia University, 1914), p. 8.

[11]Howard Galt, The Development of Chinese Educational Theory (Shanghai: The Commercial Press, 1929), p.
33.

This examination was to continue in China in various forms
for the next four thousand years.

While the Shang empire was composed of a number of
city-states, the following dynasty, the Chou (1027 B.C.-
256 B.C.) evolved into a true feudalistic society. Chou
chiefs installed themselves in Shensi and created a number
of fiefs that were governed by relatives and military
allies. In the resulting stability of this "Golden Age
of China," various institutions were to flourish.
Attempts were made at establishing a system of universal
education with schools in the capital and each feudal
state. The curriculum was extended from Shang times to
include utilitarian subjects. Studies now concerned the
six virtues, wisdom, benevolence, goodness, righteousness,
loyalty, and harmony; the six praiseworthy actions; and
the six arts, rituals, music, archery, charioteering,
writing, and mathematics. A liberal education included
five types of ritual, five kinds of music, five techniques
of archery, five ways of directing a chariot, six styles
of writing, and nine "operations" of mathematics. This
is the first mention in history of the study of mathematics in the schools of China. Pae-Yeo Yuan, in his study
of Chinese education,[12] notes that this mathematics was

---

[12]Pae-Yeo Yuan, "A Comparative Study of Chinese and
American Secondary Education" (unpublished Master's dissertation, University of Pennsylvania, 1924).

computational in nature and dealt with

1. Measurement and computation of area
2. Weighting
3. Grading merchandise
4. Computation of volume
5. Estimating building or road construction
6. Transportation
7. Finding an unknown number
8. The solution of equations
9. Work with problems involving triangles.[13]

This list appears very similar to the divisions of the
Chiu-chang Suan-shu. One, no doubt, influences the other,
but the exact direction of this influence is not clear.
Kuo outlines a scope and sequence of study for a student
of this period:

Student's Age	Nature of Studies
6-7	Learns numbers (1, 10, 100, 1000, 10,000) and the names of the points of a compass.
7-9	Respect for elders.
9-10	Taught to distinguish days, the first day of the month, the day of the full moon and the names of the days in a cycle of sixty.
10-13	Live away from home and study writing and mathematics.
13-14	Music
15	Warring arts, archery and charioteering
20	Rituals[14]

---

[13]Ibid., p. 8.

[14]Kuo, op. cit., p. 19.

Thus, although mathematics was now studied by Chinese
students, it was of the most fundamental kind. Content
was dictated by the agrarian needs of a feudalistic state.

It was during the latter part of this dynasty that
Confucius (551-479 B.C.) was born and the basic doctrines
for his school of thought formulated. Confucius, being
the product of a feudalistic society, tended to uphold
the validity of feudalism as a social institution, while
at the same time he understood that resulting injustices
had to be substantially mitigated to insure its survival.
He and his disciples incorporated into their teachings
the existing concepts of Chün-tzǔ and Hsiao-jên, or superior and inferior men, and built an ethical philosophy on
this dichotomy. For society to function harmoniously,
each segment was to assume its proper role in relation to
itself and its counterpart. Three prevalent themes
dominate the resulting Confucian thought: the dualistic
nature of society rendering one man superior, another
inferior; the duty of the superior to rule and the
inferior to be ruled; and virtue as basis for social
status. Confucius reconciled his defense of feudalism
with the dignity of the individual by specifying that
moral rectitude would provide a counterbalance for the
authoritarian character of society. Superior men had to
earn their position by acquiring benevolence, righteousness, propriety, and wisdom. This acquisition was to be

obtained through proper education. Confucian thought
would now become the dominating factor in Chinese education for the next two thousand years. Formal education
was to be ethical training, humanistic, and literary in
character, and as such would allow little opportunity
for instruction in the art and science of mathematics.

The following Han dynasty (206 B.C.-221 A.D.)
continued the practice of selecting "filial sons and
honest subjects" from a civil service-oriented examination system which had incorporated the teachings of
Confucius. This more formalized education consisted of
studying, or rather memorizing, selected classics and
developing an ability in poetry and skill in composition.[15]

The next four hundred years represent a period
of great flux in Chinese history. Political and military
intrigues fragmented the empire. Buddhist influence from
India combined with a growing Taoist movement to make
metaphysical inroads on the established Confucian philosophy. Libraries, schools, and places of worship were
destroyed. The civil service examination system partially
collapsed. Examinations were held at one place and
another sporadically, and often political consideration

---

[15]Thaddeus Wen-hsien Yang, "The Development of
Education in China" (unpublished Master's thesis, De Paul
University, St. Louis, 1951), pp. 146-147.

rather than scholarly achievement was the criterion for
the selection of worthy candidates. Although this period
witnessed social disruption, it was not a time of intellectual stagnation. Literature, including the compilation
of Confucian commentaries, progressed, and technological
and artistic advances could be noted.

Consolidation of the empire was reachieved during
the Sui and Tang dynasties (590-906 A.D.). Internal
administration was strengthened by once again requiring
officials to submit to a civil service examination that
stressed Confucian ideals. Improved fortifications and
an extended system of roads and canals assisted in unification. Official encouragement of education resulted in
an era of great intellectual activity. To assist this
process of reconstruction, the Tang dynasty instituted
new academic degrees, one of which was a degree in
mathematics, Ming-Suan, for which candidates were chosen
for their knowledge of mathematical and scientific theses.
It is recorded that in the year 742 A.D. many mathematics
degrees were available, but few candidates chose the
subject.[16] Indoctrinations of the lower classical curriculum dissuaded candidates from all but classical
pursuits. The literary nature of the examination was
already deeply ingrained.

---

[16]Kuo, op. cit., p. 45.

Under the Sung dynasty (960-1279), the system of competitive examinations reinaugurated by the Tang was developed and modified. Dissatisfaction with the strictly classical studies required of candidates and the resulting emphasis on memorization was voiced by Wang An-shih (1021-1086). Wang, a Prime Minister, attempted to use his influence in the promotion of more practical scholarly training. In his Ten Thousand Word Letter to the emperor, Sung Shen-tsung, he suggested reforms for the educational system. These were:

1. Education must accommodate individual student differences, and education for gifted children school receive special attention.

2. The curriculum should be designed to meet the needs of society.

3. The classics which did not meet the present stage of Chinese development should be abolished.

4. Poetry and rhythmical prose should be abolished from the examinations.

5. Only classics and technical knowledge were to be tested.[17]

Wang's reforms were rebuked by his fellow scholars whose positions were secured in traditional learning. Around 1104, the emperor, Hui-tsung, created four special schools: mathematics, medicine, painting, and calligraphy.[18]

---

[17]Yang, op. cit., p. 47.

[18]Kuo, op. cit., p. 45.

Their existence, however, was precarious and they soon drifted into extinction. The classical literary curriculum was withstanding all attempts to dislodge it from its position of central importance in Chinese education.

Ming emperors (1368-1644) were liberal patrons of literature and education. T'ai-tsu, founder of the dynasty, added the study of military arts and mathematics to the civil service roster.[19] These disciplines were added to the examination system at the provincial and national levels. Military qualifications included skill in archery and a demonstration of great physical strength. No indication is available as to what the mathematical requirements were, but it can safely be assumed they included a familiarity with the mathematical works then in existence and a proficiency in calculation techniques. Within a short time, education again returned to a solely literary character.

The following Ch'ing or Manchu dynasty (1644-1912) established a new series of schools, including schools for sons of officials, ordinary Manchu Bannermen, and Mongols, to teach the Manchu and Mongol language and mathematics. Two courses of study were provided by the Imperial University: the Classics and government administration. Administrative studies included such topics

----

[19]Ibid., p. 53.

as public rites, taxation, law, frontier defense, water-
ways, and mathematics. Still the classical outlook
prevailed in education. Reformers such as Yen Yuan (1635-
1704), advocate of realistic scholarly pursuits ("What I
want is motion, activity, reality, utility"),[20] and Li
Kung (1659-1733), who insisted on acquiring knowledge
based on experience and specialization in all fields of
learning, pleaded their causes to no avail. In the latter
part of the Ch'ing dynasty, as the period of strong Western
influence approached, the higher institutions of education, such as the district academies and university,
retreated into a nominal existence and functioned only at
examination time.

The Effects of the Classical Tradition on Mathematical Thinking and Instruction

By the beginning of the nineteenth century, three
thousand years of Chinese educational experience had been
compiled into an authoritarian theory of instruction.
Education was humanistic and formal. No foundations were
established to promote a scientific spirit of inquiry.
Indeed, all efforts had been directed toward suppressing
such activities.

> What is nature? What is human? That the ox and
> horse have four feet is nature; to halter the

---

[20]L. Carrington Goodrich, _A Short History of the Chinese People_ (New York: Harper and Row, Publishers, 1950), p. 227.

head of a horse or to pierce the nose of an ox
is human.[21]

Confucius' teachings urged the preservation of nature and
through it the preservation of society. Realizing that
their continued existence depended on such doctrines, the
rulers of China perpetuated Confucianism as the state
philosophy. Unfortunately, the spirit of Confucius'
original doctrines became lost in the literati's preoccu-
pation with words, calligraphic forms, and poetic styles.
Even when a "scientific" question was presented in an
examination, little theoretical knowledge or deductive
reasoning was required to answer it.

> Fire-arms began with the use of rockets in the
> Chou dynasty (1100 B.C.); in what book do we
> first meet the word for cannon? What is the dif-
> ference between the two classes of engines to
> which it is applied? [applied also to catapult]
> Is the defense of K'aifungfu its first recorded
> use?[22]

No option was left for questioning, experimentation, or
discussion. Teachers were despotic; they dictated,
students copied, then read their lesson aloud. The
volume of noise generated in such a situation was considered indicative of academic excellence. Relationship

---

[21]Chuang-tzǔ as quoted by Yu Lan, "Why China Has
No Science--An Interpretation of the History and Consequences of Chinese Philosophy," The International Journal
of Ethics (1922), 32: 237-263.

[22]W. A. P. Martin, Hanlin Papers (Shanghai: Kelly
and Walsh, 1880), p. 69.

between teacher and student is specified in the words of

the Chuang-tzŭ:[23]

> When the teacher is giving instruction, the
> pupils' proper attitude is one of docility and
> of respectful reception to the profoundest
> teaching.[24]

The only mention of mathematics in the classical

curriculum concerns very basic knowledge. Several passages

of the Trimetrical Classic[25] are devoted to introducing

the student to the fundamental concepts of the number

system:

> One is the beginning of numbers; ten is the com-
> pletion of numbers; they advance up to one hundred,
> which is a multiple of ten. Ten reduplicated and
> multiplied ten times makes one hundred; one
> hundred reduplicated and multiplied ten times
> makes a thousand; a thousand reduplicated and
> multiplied ten times makes a myriad (10,000).
> Passing this, and advancing onward, numbers are
> without any assignable limit, incapable of being
> exhausted.
>
> With regard to the fractional numbers--tenths,
> hundredths, thousandths, and tens of thousandths,
> they all decrease by tens. All numbers without
> exception are reckoned in this manner.[26]

Instruction concerning operations with numbers was left

as household knowledge and was not taught in the ordinary

----

[23]The writings of Chuang-tzŭ, a founder of Taoism,
and one of the texts of its canon.

[24]As quoted by L. G. Morgan, The Teaching of Science to the Chinese (Hong Kong: Kelly and Walsh, Ltd.,
1933), p. 53.

[25]Trimetrical Classic, a primer of basic knowledge
written in verse with stanzas of three lines each.

[26]As translated in E. C. Bridgman's Chinese
Chrestomathy (Macao: Mission Press, 1841), pp. 372-373.

school. When specific schools for mathematics were
founded, it should be remembered that these schools
existed only at the higher levels of education, and most
likely in the imperial capital alone. Mathematics was
studied in appropriate ministries of government, such as
public works or taxation, where utilitarian needed demanded
it. Then it was not studied as a science but as a collection of computational techniques necessary to achieve
specific results. The Chiu-chang Suan-shu, most widely
studied of Chinese mathematical works, reflects the needs
of a bureaucratic government. Texts such as the Chiu-
chang were published by imperial offices and distributed
within the administrative network. Seldom did they find
their way into the lower educational system or into the
hands of non-scholars. Briefly in the Sung period, there
was some mathematical activity among the common people;
but with the accession of the Ming dynasty, mathematical
studies again became confined to provincial academies.

Typical, in style and form, of the mathematics
theses studied by scholars in traditional China is the
General Comprehensive Arithmetic, compiled by an official
of the Ming dynasty.[27] This work in five volumes is a
manual of simple mathematical knowledge, including an
introduction to the number system, the four basic

---

[27]Outlined by Bridgman, pp. 372-398.

operations, mensuration, and astronomy. The exordium
alludes to the legendary beginnings of mathematics in
China:

> How did numbers originate? They originated with
> drawings and writing. Fu-Hsi obtaining them,
> drew his eight diagrams: Yu the Great obtained
> them, measured off the lands for cultivation.
> By these same means the sages arranged things in
> classes and completed their works. Of all the
> celestial officers, terrestrial magistrates,
> music, signs, soldiers, taxes, extending to the
> minutest points and the smallest fractions,
> there is not one with which numbers are not
> concerned; accordingly, there is nothing which,
> in its original principles, is not included in
> the Book of Changes.[28]

The Arithmetic is not theoretical in nature. It presents
facts, definitions, rules, and examples to reinforce them.
Many of the mathematical rules are presented in verse in
order to assist the reader in his memorization. The
decimal nature of the number system is discussed and
three systems of numeration illustrated: one for use in
books and documents, another for employment in business
accounts, and the last for use in everyday transactions.
Readers are warned, "now in reckoning numbers, we stop at
myriads [10,000] and although there are hundreds of
thousands, millions, and other [higher] numbers, they are
not very much employed."[29] Computations described are
intended to be performed on an abacus. Explanations of

---

[28]Ibid., p. 373.

[29]Ibid., p. 377.

the four operations are not elementary:

> The rule of addition is exemplified thus: if
> to each tael of money, you add interest of three
> candareens, and, without breaking the original
> sum, you increase it by three, you have the
> operation which is called addition.

> The rule of subtraction is exemplified thus:
> the capital and interest of money is four taels;
> and each tael is to be diminished by taking away
> three mace; you have then only to triple the
> three and subtract the nine, and you will obtain
> a remainder of three taels, with a fraction of
> one mace.

> The rule of multiplication is thus exemplified:
> there are four hundred and twenty-five pieces of
> cloth, and each piece costs two mace and five
> candareens; it is asked, what will be the amount
> of the money required for the whole [of said
> cloth]?
> The answer is, one hundred and six taels, two
> mace, and five candareens.
> The operation is performed thus: the cloth is
> taken as the multiplicand, and the price of each
> piece of cloth, two mace and five candareens, as
> the multiplier; then by multiplying them
> together, the product is obtained.[30]

A similar explanation is given for division. For their

basic operation facts, students relied upon the use of

tables until the facts were memorized. Work on mensura-

tion of length, weight, time, and area was also considered

in this text. Formulae for computing the areas of

specific plane figures were contained within the solution-

explanations of problems:

> Question: What is the area of a right-angled
> triangular piece of ground, whose longest side
> is sixty pu and has a base of thirty pu?

---

[30]Ibid., pp. 378-380.

Answer: Three mou, seven fan and five li [3.75 acres]

Rule: Take the perpendicular, sixty pu, and the base, thirty pu; multiply them together, and a product will be obtained, one thousand eight hundred pu; take away one half, and there are nine hundred pu, which being reduced according to the land measure, the answer is obtained.[31]

Such was the mathematics instruction presented to a scholar during his higher course of study. For the common man, artisan or peasant, no such systemized course of instruction was available. Each craft devised its own system of measure and computation, a practice which in later times would hinder a scientific unification of China. Bridgman's writings in the first half of the nineteenth century note that the ch'ih (Chinese foot) as fixed by the Mathematical Board at Peking was 13.125 English inches; that used by tradesmen in Canton varied from 14.625 to 14.81 inches; public works engineers used a 12.7 inch foot; and that which measured distances was approximately 12.1 inches.[32] Other observers of the Chinese scene were to comment on the "spirit of inaccuracy" in China and the use of approximations in lieu of exact answers which resulted from such practices.[33]

This was the legacy that modern education would be built upon. Highly formalized studies had obscured

---

[31]Ibid., p. 386.

[32]Ibid., p. 381.

[33]Morgan, op. cit., p. 55.

the principles intended for transmission and replaced them with stylistic rhetoric. The classical conception of education precluded the study of science, including mathematics, and created a dichotomy of knowledge: that which was appropriate for a scholar to study and that which was not. Mathematics and science were stigmatized as inappropriate for intellectual pursuit. The state philosophies of Confucianism, Taoism, and Buddhism reflected these views. Buddhism, in its opposition to the scientific study of nature, tied men to the past. Taoism advocated knowledge by intuition:

> Without going out of the gate one may know all under the skies. Without looking from the window, one may view the heavenly Way. The farther one goes in travel, the less he knows. Therefore the sage knows without activity and discriminates without having seen.[34]

Confucius' main contention was that man need not concern himself with that which lay beyond him. L. G. Morgan, in his thesis, The Teaching of Science to the Chinese, lists Confucian obstacles to the development of science in China:

1. While Confucian teaching itself was not antagonistic to science, it placed so much stress on human behavior and relations that material and non-human phenomena were necessarily greatly neglected.

2. Confucius' approval of the Book of Changes as a classic containing ultimate truth

---

[34]Tao-te-ching, as quoted by Morgan, p. 30.

assisted in maintaining and spreading mystical
belief by encouraging divination and the Yin-
Yang system, and thus militated against
investigation.

3. That the detailed commentary work which
resulted from the reverence for the Confucian
Classics did much to stultify intellectual
thought and limit it to literary activities.[35]

As a result of such influences, methods of scientific
inquiry never had a chance to adequately develop in China.
Mathematics, while appreciated for its utilitarian value,
was never developed into a formal system of knowledge,
nor were efforts expended to teach it as such.

The Advent of Mathematics Education in Modern China

By the beginning of the nineteenth century, the
Ch'ing dynasty was in a state of rapid decline. Incompe-
tent and corrupt internal administration had resulted in
a lack of confidence in the officialdom. Exactions of
absentee landlords, the breakdown of irrigational facilities, and natural calamities fomented a state of internal
unrest. An ensuing series of bloody conflicts, the Opium
War of 1839-42, the Anglo-French War of 1856-60, and the
T'ai-ping rebellion further weakened social institutions
that had survived for four thousand years. China's
ethnocentric cultural tradition could not support mounted

[35]Ibid., p. 28.

archers nor banner-decked war junks against the ordinance and gunboats of foreign expeditions. The "Middle Kingdom," in its self-ordained position between Heaven and the barbarian hordes, now found itself forced into humiliations and compromises. Concession after concession, including the opening of ports for foreign trade, extraterritorial rights for foreign nationals, and the establishment of a maritime customs service headed by an Occidental, was granted by the Chinese.

Finally, it became evident to the officials of China that the survival of their kingdom rested in securing a modern military force for self defense. It became painfully obvious to the Chinese that the effectiveness of any nineteenth-century military organization rested upon the support of a scientific technology, which at that time they did not possess. In order to acquire this technology, reformers at the Imperial court urged a revision of the traditional education system to include practical and scientific studies. Paramount among the new disciplines to be studied was mathematics, as it was felt to provide a foundation for all scientific work. Although the conservatives' opposition to scientific studies remained strong, three significant events promoted a general acceptance of mathematics as a subject worthy of study:

    1.   The scholar-general Tsêng's patronage of
        the printing of a complete translation

of Euclid's Elements of Geometry.

2. The establishment of an Imperial College which admitted mathematics in its curriculum.

3. The placement of mathematical knowledge on an equal state priority with classical studies by its inclusion in the civil service examination of 1887.

These events, together with an increased popular exposure to Western mathematical literature, nurtured an intellectual climate favorable to mathematics education.

The Completion of the Translation of Euclid's Elements of Geometry into Chinese

Most prominent among the officials of nineteenth-century China who urged educational reforms was Tsêng Kuo-fan. Tsêng, a scholar-general instrumental in subduing the T'ai-p'ing revolt, was accidentally recruited into the reform movement. Near the close of his campaign against the rebels, he was approached by the scholar Li Shan-lan, who worked under him in the Anking army. Li requested Tsêng to sponsor the printing of a book which was "indispensable to mathematicians" and which would be "lost if not printed at once." This book was a completed translation of Euclid's Elements. While Hsu Kwang-shi of the Ming dynasty had transcribed Matteo Ricci's dictates of the first six chapters of the Elements, the work remained uncompleted for centuries. In 1851 Li Shan-lan joined with the Western scholar, Alexander

Wylie,[36] to finish a translation of the last nine chapters. Their work was published by a Lu-ching of Sungkiang; but soon after the wooden printing blocks were burned in the course of the rebellion. Now Li sought Tsêng's assistance in reprinting this book.

During the clashes of the T'ai-p'ing revolt, Tsêng had witnessed the effectiveness of the foreign mercenaries' war machines, and developed an admiration for the technology that could produce them. He became an active supporter of the study of Western science in Chinese schools as it immediately related to the art of war. No doubt as he read Ricci's preface to the first Chinese edition of the Elements, his suspicions concerning the importance of mathematics were confirmed. Ricci distinguished between arithmetic, a practical science, as compared with geometry, a more theoretical one, and indicated the importance of geometry to Chinese society in the fields of astronomy, the making of calendars, medicine, geography, and commerce. In particular, he stressed the need for geometry in the conduct of military affairs:

> Military affairs are the most important of all
> in the nation, for upon them its security
> depends. . . . Therefore, a general of wisdom
> and bravery must first of all have a knowledge

---

[36]Wylie had established a translation bureau at the London Mission Hospital in Canton.

of geometry, otherwise his knowledge and bravery
will be without practical value.[37]

An example was then cited of how a general of a small
nation conquered a larger one through the use of his
knowledge of geometry.

Tsêng ordered a revised edition of the completed
Euclid's Elements of Geometry to be published from his
headquarters in Nanking in 1865. To the previous prefaces
contained in the translations written by Ricci, Hsu, and
Wylie and Li, Tseng added his own in which he presented
his opinion concerning past mathematics education in
China:

> According to our traditional mathematics, each
> section derives its name from a specific
> (practical) function. The students all follow
> rules in solving their problems. All their
> lives they use mathematics knowing only how to
> do it and not why it is done. Therefore, they
> consider mathematics as a very difficult subject
> simply because they are confused, knowing the
> method, but not the principle. . . . Euclid's
> Geometry (on the other hand) deals not with
> method, but with principles.[38]

In the intervening two centuries since Matteo
Ricci first dictated his geometry, other missionaries
translated scientific works into Chinese; however, only
twelve concerned mathematics and none completed Euclid.

---

[37]Cyrus Peake, Nationalism and Education in Modern
China (New York: Columbia University Press, 1932), p.
7.

[38]Ibid., p. 6.

Now at a most critical period of Chinese intellectual
indecision, Tsêng, a renowned official of the empire,
sanctioned the publication of the most influential of
all Western mathematical works. This action was responsible for increasing the advocates for reform and, in
particular, advocates for mathematical studies.

Formation of the T'ung-wên-kuan

Among the concessions exacted from China in the
mid-nineteenth century was a pledge to conform to a
theory of international relations founded on a concept
of equal and sovereign states.[39] To facilitate this
pledge, a foreign office, the Tsung-li Ya-mên, was founded
in Peking and placed under the direction of a prince of
the empire, I-hsin (1833-1898). Commonly known as Prince
Kung, I-hsin was to use this office in promoting radical
reforms. One of the innovations instituted under the
influence of the Tsung-li Ya-mên was the development of
an Imperial Interpreters College, eventually to be known
as the T'ung-wên-kuan or School of Combined Learning.

The plan for this school was put forth in a memorial of January 13, 1861. In this document, Kung and
the scholars Wen-hsiang and Kuei-liang advised that the
key to Western knowledge was to be found in the learning

---

[39]Tientsin Treaties, 1858, and the Peking Treaties,
1860.

of the barbarians' language and the translation of their books. The memorial was granted, and by August of 1862, the school was functioning with its first classes in English. Later, the languages of French, Russian, and Japanese were also added to the curriculum. Fêng Kuei-fen, a Soochow scholar and experienced government administrator, agreed with the concepts of this new school, but urged an extended curriculum:

> At the same time they [the students] should learn mathematics. All Western knowledge is derived from mathematics. Every Westerner of ten years of age or more studies mathematics. If we now wish to adopt Western knowledge, naturally we cannot but learn mathematics . . .[40]

In June 1863, Li Hung-chang, a well-known administrator, supported the establishment of the T'ung-wên-kuan and urged that a similar school be opened in Shanghai. His eloquent memorial was drafted by his secretary, Feng Kuei-fen, and combined both men's sentiments concerning the necessity of mathematical studies:

> I have learned that when Western scholars make weapons, they use mathematics for reference and exert their energy in deep thinking to make daily increases and alterations.[41]

In response to this memorial, similar language schools were established in Shanghai in 1863; Canton, 1864; and

---

[40] John K. Fairbank and Ssu-Teng, *China's Response to the West: A Documentary Survey* (Cambridge, Mass.: Harvard University Press, 1954), p. 51

[41] *Ibid.*, p. 70

Foochow, 1866. The inception of these schools initiated
the government's first attempts at modern education.

On December 11, 1866, Prince Kung voiced his argu-
ment for the extension of the T'ung-wên-kuan's studies to
include mathematics and science:

> The machinery of the West, its steamers, its
> firearms, and its military tactics, all have
> their source in mathematical science. Now at
> Shanghai and elsewhere the building of steamers
> has been commenced; but we fear that if we are
> content with a superficial knowledge and do not
> go to the root of the matter, such efforts will
> not issue in solid success . . . For we are con-
> vinced that if we are able to master the
> mysteries of mathematical calculations, physical
> investigations, astronomical observation, the
> construction of engines, the engineering of
> waterways, this and only this will assure the
> steady growth of the power of the empire.[42]

Again on January 28, 1867, Kung petitioned for
scientific studies. In this more detailed memorial, he
played upon the chauvinistic pride of the court by
explaining that this learning would not be the acquisition
of barbarian knowledge because Western science had really
borrowed its roots from ancient Chinese mathematics. Two
heroes of the empire were held up as patrons of mathemat-
ics--Emperor K'ang-hsü (1662-1723), who encouraged the
study of mathematics during his reign, and Tsêng Kuo-fan,
who had just published a completed translation of Euclid's
Elements. Conservatives in the court were quick to rally

---

[42]Translated in W. A. P. Martin, A Cycle of Cathay
(London: Olephant, Anderson and Ferrier, 1900), p. 301.

against the implications of these memorials. Their

spokesman, Grand Secretary and head of the Han-lin Acad-

emy, Wo-jin, cautioned against the dangers of such an

action:

> Mathematics, one of the six arts, should indeed
> be learned by scholars as indicated in the Impe-
> rial decree, and it should not be considered an
> unworthy subject. But according to the viewpoint
> of your slave, astronomy and mathematics are of
> very little use. If these subjects are going to
> be taught by Westerners as regular studies, the
> danger will be great . . . Your slave has learned
> that the way to establish a nation is to lay
> emphasis on propriety and righteousness, not on
> power and plotting . . . If astronomy and mathe-
> matics have to be taught, an extensive search
> should find someone who has mastered the tech-
> nique. Why is it limited to barbarians, and why
> is it necessary to learn from the barbarians?[43]

Prince Kung responded with a rebuttal from the

Tsung-li Ya-mên. Kung recalled the past humiliations

suffered at the hands of the foreigners. He vividly

described how, when the capital was in peril, the "schol-

ars and officials either stood about, putting their hands

in their sleeves, or fled in confusion."[44] Then he

expounded on the secret of the foreigners' power:

> Moreover, the principal means which foreigners
> employ to secure victory is the use of steam-
> ships and firearms first of all. Formerly,
> because the Europeans, in making firearms, did
> not care how much capital they used and because

---

[43]Fairbank and Teng, op. cit., p. 76.

[44]Paul Clyde and Burton Beers, The Far East (Englewood Cliffs, N. J.: Prentice Hall, 1966).

firearms have their roots in astronomy and
geometry and are developed by trigonometry and
mathematics, so that their guns can be cleverly
discharged and marvelously hit the mark, the
censor Wei Mu-t'ing requested that at Shanghai
and other spots factories be established . . .
Your ministers have also discussed this in correspondence with Tsêng Kuo-fan, Li Hung-chang,
Tso Tsung-t'ang, Ying-kuei, Kuo Sung-tao,
Chiang I-li, and others. They all agree that
the clever methods for manufacturing must begin
with mathematics . . .[45]

An imperial edict of March 5 rejected Chang's

criticisms, ". . . it is incumbent on our scholars to

learn the rudiments of mathematics and astronomy,"[46] and

on July 3rd the Tsung-li Ya-mên announced the results of

an entrance examination for mathematics students; of

seventy-two applicants, thirty were accepted. The cur-

riculum and faculty were expanded to accommodate the new

disciplines. By 1869, the faculty totaled thirteen

professors, four Chinese and nine Europeans. The Profes-

sor of Mathematics who was credited with "awakening a

taste for such studies in the minds of his countrymen"[47]

was Li Shan-lan, the co-translator of Euclid. Enrollment

rose to 120 students and the regular course of study

developed into an eight-year program, with Chinese

---

[45]Fairbank and Teng, op. cit., p. 76.

[46]Translated in Meribeth C. Cameron, The Reform
Movement in China (New York: Octagon Books, Inc., 1963),
p. 18.

[47]Martin, op. cit., p. 310.

in the morning and Western studies in the afternoon. This "combined learning" was distributed as follows:

Year

I   Reading, writing and composition of simple sentences, oral presentation of readings from simple books.

II   Continuation of oral readings, exercises in sentence structure, simple translations.

III   World geography, world history, translation of selected essays.

IV   Arithmetic, algebra, translation of official documents.

V   Natural science, geometry, plane trigonometry and translation of books.

VI   Study of machinery, calculus, nautical surveying and translation.

VII   Chemistry, astronomy, international law, translation.

VIII   Astronomy, surveying, geography, finance, translation.[48]

For students not involved in language, a shorter five-year course was available. Its stress on mathematics is obvious:

Year

I   Arithmetic, algebra, and Chinese mathematics.

II   Geometry, plane and spherical trigonometry.

III   Natural philosophy, chemistry.

---

[48]Knight Biggerstaff, "The T'ung Wen Kuan," The Chinese Social and Political Review (1934), 18: 239.

IV   Differential and integral calculus, naviga-
      tion and surveying, and theoretical and
      practical mechanics.

V   International law, political economy,
      astronomy, geology, and mineralogy.[49]

This curriculum provides the first instances of
formal mathematics instruction in the modern Chinese educational system. While there is no indication of the
level of rigor at which mathematics was taught, it would
probably be safe to assume that Li Shan-lan would use his
translations of Euclid, De Morgan, and Loomis. The
pioneering accomplishments of the Tsung-li Ya-mên
instigated further reforms directed at the translation of
Western books and the securing of practical knowledge.

Mathematics in the Traditional Examination System

Establishment of modern schools with curricula that
included mathematics and science exerted a minimal impact
on the traditional education system. The system was still
concerned with producing scholars whose training would
allow them "to get the right word at pencil's point or
tongue's end." As long as the civil service examination
required scholars trained solely in the knowledge of the
Confucian Classics, the efficacy of scientific studies
would be limited. The imprimatur to sanction the study

---

[49] Ibid.

of science and mathematics as a dignified discipline
worthy of scholarly study was to be the inclusion of
questions concerning these subjects on the state examinations. In 1869, the Viceroy of Fukien province memorialized the throne asking that mathematics be admitted in
the examination system.[50] Without waiting for official
sanction, two provincial superintendents of education
made attempts to introduce mathematics on their examinations. In 1874, there was a mathematics examiner present
at the Hunan examination, but no candidates chose the
subject; and in 1885, the Shantung examination requested
and received a few mathematics papers.[51] Li Hung-chang,
Viceroy of Chihli, again formally requested that mathematics and science be admitted to the examination. When
Martin discussed the possibilities of such a reform with
Grand Secretary Shên Kuei-fên, he was told that the reform
would be accomplished, and as a result students would
seek out science masters to study under; the government
would not have to provide schools.[52]

Under continuous pressure, the throne finally
conceded and decreed in 1887 that mathematics questions
be included in the state examination. The triennial

[50]Kuo, op. cit., p. 66.

[51]Martin, op. cit., p. 319.

[52]Ibid.

examination at Peking in 1888 saw the first instance of
this new policy. Of the sixty candidates present, thirty-
two were allowed to take the test and one emerged successful. As the state's need for trained mathematicians
became more obvious, Shên Kuei-fên's prediction was proven
false. An edict of the emperor, endorsed by the Tsung-li
Ya-mên in 1896, commanded that foreign mathematics and
sciences be taught in all Yamens of the empire. From
then on, every candidate for the civil service examinations would be expected to qualify in at least one
science and in mathematics.

The Publishing of Modern Mathematical Works

As the Chinese awareness of Occidental accomplishments grew, the demand for scientific literature, both
indigenous and Western, increased. The need for mathematics books was satisfied by three sources: official
government translation agencies; foreign missionaries
compiling materials for their schools; and native writers
concerned with China's contemporary intellectual needs.

Alexander Wylie, A British scholar, officially
invited to China as a translator, established himself at
the London Mission Hospital in Canton. There he was
joined by the Chinese scholar, Li Shan-lan, and by the
Reverend J. Edkins. Together these men translated many
Western works, including Euclid's <u>Geometry</u>, Whewell's

Mechanics, Hershel's Astronomy, De Morgan's Algebra, and Loomis' Conic Sections and Infinitesimal Calculus.[53]

The Tung-wên-kuan united with arsenal translation bureaus at Shanghai, Foochow, and Kiangnan to turn out Chinese versions of Western tracts. Kiangnan's first publication was Practical Geometry, translated and compiled by John Fryer and Hsu-jun. By 1880, the Kiangnan bureau alone had translated ninety-eight works, of which twenty-two were about mathematics.[54] While the majority of these books were written for advanced study, the bureau also translated a forty-three volume school textbook series which included works about mathematics. Products of Kiangnan Translation Bureau were circulated by imperial order to all provincial yamens, and provided fuel for the intellectual revolution sweeping China.

The arrival of missionaries in China followed closely in the wake of diplomats and merchants. Almost

---

[53]William Whewell, The Mechanical Euclid (Cambridge: Cambridge University Press, 1843); John Herschel, Outlines of Astronomy (London: Longman, Green and Roberts, 1859); August De Morgan, Elements of Algebra Preliminary to the Differential Calculus (London: Taylor and Watson, 1837); Elias Loomis, Elements of Geometry and Conic Sections and Elements of Differential and Integral Calculus (New York: Harper and Brothers, 1847, 1874).

[54]John Fryer, "Science in China," Nature (May 19, 1881), p. 55.

all of these evangelists founded schools. Although the
curricula of missionary institutions were mainly humanistic and designed for the purposes of proselytizing,
they did include mathematical and scientific studies.
The Rev. Charles Mateer, teaching at his boys' school in
Tengchow, Shantung, noticed his pupils' disdain for
mathematical computation on the abacus. In classical
scholarly fashion, they scorned such knowledge as befit-
ting shopkeepers and merchants. Mateer decided to teach
mathematics the "Western way" and used a small arithmetic
published by Gibson of Foochow.[55] When this work went
out of print, Reverend Mateer wrote his own, which was
published in 1877 in two small volumes. This book was
widely circulated throughout the empire. Mateer's
Arithmetic introduced two new concepts into popular
Chinese mathematics: Arabic numerals and increasing
positional values of digits varying horizontally rather
than vertically as in the Chinese fashion. In the explanation of this procedure, problems would appear in pairs,
one in traditional form, the other in Western notation:

---

[55]An Occidental employed at the translation bureau
in Foochow.

```
    Example                          Again

1
3       3   4   3                      3896
1   4   8   8   8                      4894
2   3   9   9   9                      3894
1   7   4   4   6                       437
                                     ─────
                                     13121  56
```

Although his students called this "crab writing" because
it walked sideways, they adapted to it quickly. Mateer
compiled more advanced texts as they were needed. He
published a book on geometry in 1885 and part I of an
algebra text in 1888; the second part would follow in
1908. A colleague of Mateer's, Junius H. Judson, teaching at the Hangchow Presbyterian Boy's School, wrote a
popular book on conic sections.

As a reaction to the influx of foreign mathematical material, native authors produced revisions of
classical mathematics texts. Three books of this type
to appear were: the Shou-shan-ko Ts'ung shu of Matteo
Ricci, 1884; the Chao-tai Ts'ung-shu, containing the
works of John Schaals,[57] 1883; and the Great Treasure
House of Chinese and European Mathematics,[58] 1889. This

---

[56]Charles H. Corbett, Shantung Christian University
(New York: United Board for Christian Colleges in China,
1955), p. 15.

[57]John Schaals, a Jesuit Missionary who followed
after Ricci.

[58]For a detailed discussion of this work, see Louis
Vanhee, "The Great Treasure House of Chinese and European
Mathematics," The American Mathematical Monthly (1926),
33: 502-506.

last work, edited by Ch'en Wei-ki, contained one hundred sections encompassing writings from the Chou-pei through the contributions of seventeenth- and eighteenth-century Jesuit scholars to the recently translated works on calculus and conic sections.

Conclusions

Although China experienced mathematical activity at a very remote date, the knowledge obtained was not refined nor developed into a science. Economic conditions in early China gave rise to a social and moral code that precluded the applications of science to ease human conditions. While philosophical speculation was permitted, it functioned within the constrains of accepted servitude to nature, the spirits and the emperor. Confucian, Taoist and Buddhist doctrines reaffirmed the principle that the preservation of Chinese society was secured in a humanistic morality, one devoid of scientific conjecture. In such an intellectual climate mathematical studies, even at a very elementary level, were almost nonexistent.

With the impact of the West upon China in the early nineteenth century, conditions for social and cultural survival were reversed. Lacking modern industry and armaments, China found herself at the mercy of foreign invaders. In an attempt to remedy the situation, scholars sought out the "secrets" of Western machines. Thus the need for science-oriented education became apparent to some officials of the Middle Kingdom. These officials, led by Prince Kung, felt that mathematical studies were of primary importance and campaigned for the inclusion of this discipline into the classical curriculum. The years between 1860 to 1900 mark a long and often bitter period of controversy over this concession. Court conservatives, aware of their vulnerability, argued that such knowledge was alien to Chinese experience. Effects of their opposition were diminished by a series of timely events which included: a massive translation of Western scientific works, among which was a complete edition of Euclid's Elements; the establishment of mission schools with modern curricula and the founding of an imperial college, the T'ung-wên-kuan, with mathematical studies in its program. Finally, with the admission of mathematics questions on the Traditional Civil Service Examination in 1888, mathematics became officially accepted as a subject worthy of study. Despite this acceptance, the implementation of broad mathematics education reforms would be difficult to achieve. China entered the twentieth century committed to the establishment of a national school system with a modern curriculum but the lingering prejudices of her traditional past would impede this transformation.

CHAPTER II

MATHEMATICS IN THE MODERN SCHOOL SYSTEM,
1903-1949: THE EMERGENCE OF PROBLEMS

Under the Manchu Government, 1903-1912

The turn of the century found China irrevocably
committed to educational reform. Shortly after her
return to the throne, the Empress Dowager expressed her
convictions concerning reform:

> The board of Rites have further asked that I put
> a stop to the new studies ordered to be taught
> in the provincial colleges, and newly instituted
> modern schools, and that these institutions be
> commanded to return to the old regulations--that
> of giving instructions solely in the Confucian
> Analects and ancient classics. Now the objects
> [objectives] of these colleges as institutions
> of learning is to teach solid and substantial
> branches of study and not solely old time books.
> For instance, such subjects as astronomy,
> geography, military strategy, mathematics and
> such are modern requirements necessary for the
> country's welfare, and as such within the
> province of what educated students should
> acquire.[1]

Now, even the conservatives conceded the need for change.
In 1901, Sun Chia-nan, President of the Han-lin Academy,
memorialized the throne requesting that members of the
Academy be required to study mathematics and science

---

[1]North China Herald (October 30, 1899), p. 874.

rather than calligraphy and poetry.[2] In the same year,
the Literary Chancellor of Shantung province prepared a
study list of modern knowledge for prospective Hsi Tsairs
in which mathematics was included. Alteration of the
civil service examination was not enough. The examination
system itself was under attack and verged on extinction.
Desired reforms could only be realized through the establishment of a modern educational system!

Formation of the Modern School System

July and August of 1901 saw the beginning of a
series of royal edicts that would eventually culminate in
the establishment of a modern school system. On October
10, the outline of a national school system was presented
to the throne. Its close resemblance to the Japanese
system was rationalized by the Board of State Affairs:

> Japan is of the same continent with ourselves;
> her change of methods is quite recent, and she
> has attained to strength and prosperity. Her
> experience has been so nearly like our own that
> we may derive instructions from it.[3]

The system provided for kindergartens and primary and
secondary schools with curricula designed to accommodate
the new civil service examination requirements. Execution
of this scheme in the districts and provinces met with

---

[2]Ping-Wen Kuo, op. cit., p. 75.

[3]United States Consular Reports, Washington, D. C.,
January 1902, p. 30.

varied degrees of success; teachers, books, and finances for such an undertaking were lacking. In 1902, Chang Po-hsi, Chancellor of the Imperial University, drafted a general revision of the system; however, Chang's revision was not accepted. Joining with the educators Chang Chih-tung and Yung Ching, he drafted another more detailed set of regulations for the new schools. This new system advocated

1. Loyalty to the monarch
2. Reverence to Confucius
3. Promotion of the public spirit
4. Promotion of a martial spirit
5. Utility[4]

It was presented to the throne on January 13, 1904, and accepted. In execution of these ideals, the Japanese system was still closely followed. Kindergarten preceded five years of lower primary school, which was followed by four years of upper primary, five years of middle school, five years of higher school (college), and three or four years of university study. Five years of graduate studies ended the sequence.

---

[4]Yang, op. cit., p. 53.

TABLE 1

PRIMARY SYLLABUS FOR MANCHU SCHOOLS[5]

Lower Primary		Upper Primary	
Subject	Hours/week	Subject	Hours/week
Chinese Classics	12	Morals	2
Ethics	2	Chinese Literature	8
Chinese Literature	4	Chinese Classics	12
Mathematics	6	Mathematics	3
History	1	Science	2
Geography	1	Chinese History	2
Science	1	Geography	2
Physical Exercise	3	Drawing	2
		Physical Exercise	3
	30		36

Lower primary studies consisted of eight subjects comprising thirty hours over the six-day school week (see Table 1). An official outline of the mathematics studies is very brief:

Year	Activities
1	Reading and writing of the numbers 1-20, simple addition and subtraction.
2	Reading and writing of numbers 20-100, introduction of the operations of multiplication and division.
3	Work on the four operations
4	Introduction of decimals and use of the abacus.

---

[5]H. E. King, The Educational System of China as Recently Constructed, Bulletin 1911, No. 15 (Washington, D. C.: Government Printing Office, 1911), p. 51.

5   Practice on the four operations
using an abacus.[6]

Upper primary extended the curriculum to nine
subjects and the school week to thirty-six hours. Mathematics studies completed elementary arithmetic and
provided more practice on the abacus.

The middle school curriculum increased the number
of different subjects studied to twelve, while still providing thirty-six recitations a week. Mathematics studied
at this level included advanced arithmetic, algebra,
geometry, and plane trigonometry. Table 2 gives the full
middle school curriculum.

TABLE 2

MANCHU MIDDLE SCHOOL SYLLABUS [7]

Subject	_____Hours/Week_____					
Year:	1	2	3	4	5	Total
Self Culture	1	1	1	1	1	5
Chinese Classics	9	9	9	9	9	45
Chinese Literature	4	4	5	3	3	19
Foreign Language (English)	8	8	8	6	6	36
History	3	2	2	2	2	11
Geography	2	3	2	2	2	11
Mathematics	4	4	4	4	4	20
Nature Study	2	2	2	2	–	8
Drawing	1	1	1	1	–	4
Physical Science	–	–	–	4	4	8
Military Drill	2	2	2	2	2	10
Option	–	–	–	–	3	3

---

[6]Ibid.

[7]Jen-Mai Tan, "History of Modern Chinese Secondary
Education" (unpublished Ed.D dissertation, University of
Pennsylvania, 1940), p. 78.

Implementation of this program was assisted by a royal decree of September 1905 banning the civil service examination completely. Now, student hesitancy in accepting modern studies instead of a classical curriculum became unwarranted. Even in this climate of acceptance, the accomplishments of the new schools soon became questionable. Standards were not uniform and, when existent, were extremely low. Many schools did not have pupils prepared for the level of instruction they were offering. The newly formed Ministry of Education memorialized the throne in 1906 criticizing a lack of clarity in the then existing educational aims and presented five points it believed should be adopted. Of this list, the fifth point stressed the need for utility in the curriculum:

> The textbooks in Chinese schools should hereafter place less emphasis on theory and more on facts. Scientific research, drawing and handiwork should be added to the curricula in order to fit the children for industrial pursuits.[8]

School conditions were investigated in 1909; hardly one school was found adhering to government standards.[9] On May 15, 1909, the Ministry requested that the courses of study be altered and proposed dual-track programs for both primary and secondary schools. One track of primary school extended the school week to seven days (36 hours),

---

[8]Peake, op. cit., p. 66.

[9]King, op. cit., p. 57.

with Sunday mornings used for reviewing the week's work.
This track combined history, geography, and natural science
to form "general knowledge." Music and drawing were added
to the program. The second track suggested an easy course
of reading, Chinese language, literature, and mathematics.
Similarly, the secondary program was divided into two
streams--arts and science. Each department had major and
minor subjects. The minor subjects for the art department
were ethics, mathematics, biology, physics, chemistry,
economics, political science, drawing, and physical training. For the science department they were ethics,
classics, literature, history, geography, drawing, manual
training, economics, political science, and physical drill.
For a listing of the major subjects in this program see
Table 3 below.

TABLE 3

MAJOR SUBJECT IN DUAL-TRACK MIDDLE SCHOOL
PROGRAM, 1909 (HOURS/WEEK)[10]

Arts	Year:	1	2	3	4	5
Classics		10	10	10	10	10
Foreign Language		6	6	6	6	6
History		3	3	3	3	3
Geography		3	3	2	2	2
Science						
Foreign Language		10	10	8	8	8
Mathematics		6	6	6	6	6
Biology		6	6	–	–	–
Physics		–	–	8	–	–
Chemistry		–	–	–	8	8

[10]T. T. Teng and T. T. Lew, Education in China
(Peking: The Society for the Study of International
Education, 1923), p. 11.

This educational system not only paralleled the Japanese system in its conception, but also in its execution. Japanese influence permeated the content and methods of instruction. This influence was perpetuated by the employment of large numbers of Japanese teachers in the schools, massive translation and adoption of Japanese texts, and the normal training of thousands of Chinese teachers in Japan. Thus, the adopted educational system was in many respects alien to the Chinese experience.

Mathematical Content in the Curricula

In the February 1909 issue of Educational Review, Professor Chester Fresson of the Canton Christian College described an idealized mathematics curriculum for a mission school. While his program diverges from government standards, it is interesting in that it reflects the opinion of a foreign observer as to the mathematical competence of Chinese students. His scope and sequence covered six years of a seven-year program.[11]

In the first two years of instruction, Fresson advocated no formal arithmetic teaching. A concrete approach employing sorting and ordering activities

---

[11]Chester J. Fresson, "The Course of Study in the Mission School," Educational Review (Educational Association of China, Shanghai, 1909), 2: 1-12.

instilled concepts of quantity, size, weight, and shape.
Drawing and model construction were to be employed in
introducing the students to geometrical concepts. During
the third year of school, the four operations with any
combinations of numbers were considered. Work on combined
operations was done. An integrated approach to teaching
was still used, encouraging an extension of knowledge concerning geometrical figures and the introduction of
algebraic symbols. Student work was to be, to a large
extent, oral, with emphasis on speed and accuracy in
computation. The fourth year stressed the logical presentation of problems in written and oral form. Algebraic
and geometric concepts were to be directed at the solution
of practical problems. In the fifth year more difficult
problems were considered, including work with rates and
scale drawings. The last year of this sequence was
devoted to advanced arithmetic, including percentage
computations and "business arithmetic." While preparing
the student for higher mathematical studies, the program
also considered the needs of the terminal student. The
time allowed for this course was two hours a week during
the first year, two-and-a-half during the second through
the fifth, and five hours a week in the last year.

A scope and sequence published in a later issue of
the Educational Review adheres more closely to official
standards. This program was followed at the West China

Union Primary School in 1911:

Lower Primary	Topics Covered
I	1. Counting of numbers to 50 using objects. 2. Writing of numbers to 50. 3. Addition facts to 10 using objects. 4. Oral problems with mixed combinations. 5. Counting by two's to twenty and by three's to eighteen.
II	1. Review work of previous grade-free use of addition facts to 10. 2. Counting to 100 with and without objects. 3. Subtraction combinations to 10, taught with objects. 4. Oral problems concerning mixed combinations. 5. Counting by five's and ten's to 50. 6. The writing of numbers to 100.
III	1. Review and drill of previous work. 2. Addition combinations to 20. 3. Writing and counting of numbers to 200. 4. Counting by two's to 50 and three's to 60. 5. Practice oral problems.
IV	1. Review of previous work. 2. Introduction to the operation of multiplication. 3. Counting to 500. 4. Counting by five's and ten's to 100. 5. Written problems involving addition and subtraction.
V	1. Review the three operations. 2. Development of multiplication facts to five using objects. 3. Introduction to division. 4. Drill in rapid counting.

Lower Primary (Continued)

5. Writing of numbers.
6. Solution of pupil constructed problems.

Senior Primary School

I

1. Review, provide strong drill on multiplication and division without exceeding three digit numbers.
2. Develop the method of determining any two factors of numbers at least to sixty by inspection.
3. Work with compound rules, weights, and measures. Preparation for fractions by oral instruction.

II

1. Review multiplication and division using four-digit numbers.
2. Finding G.C.D. and L.C.M. by inspection.
3. Reduction, addition, subtraction, and multiplication of common fractions with the denominator to 36.
4. Simple decimals.
5. Reduction of denominate numbers.

III

1. Review denominate numbers.
2. Teach factoring thoroughly, and distinctive features of multiplication and division, finding G.C.D. and G.C.M.
3. Teach common fractions.
4. Teach operations on decimals using Szechwan dollars.
5. Commercial arithmetic, including profit and loss problems, taxes, insurance and interest.

IV

1. Percentage, ratio, and proportion.
2. Techniques of finding square and cube roots.

Senior Primary (Continued)

      3. Mensuration of plane
         surfaces.
      4. Properties of polygons,
         trapezum, cone, sphere,
         and cylinder.[12]

    Texts: Arithmetic, Mateer
            Advanced Arithmetic, Liu Gwang
            Dzao, The Commercial Press, Shanghai

Two unusual features of this description stand out: the
late introduction of common fractions and decimal fractions into the course, and the apparent lack of any
computation employing an abacus. The Chinese, in following the examples of their Japanese mentors, reasoned that
children had little difficulty with decimals and little
need of fractions.[13] These topics were put aside to make
room for more important content. Although utilization
of the abacus in schools was indicated in the government's outline, there appears to have been a social stigma
associated with its use making it "unbecoming" a scholar.
Whether due to facility or haughtiness, Western methods
of computation, as preferred by the Japanese, employing
a pencil or writing brush and paper, were followed.

    While official information concerning the mathe-
matics taught at the secondary level is not available, it

---

    [12]"West China Union Course of Study," Educational
Review (Shanghai, 1911), 4: 12-21.

    [13]Baron Kikuchi, Japanese Education (London: John
Murray, Albemarle St., 1909), p. 177.

would be difficult to imagine the content differing
greatly from that specified in contemporary Japanese
syllabuses. If this was indeed the case, the program
would have consisted of four hours a week devoted to the
study of arithmetic in the first year and two hours in
the second year.[14] The first year's work reviewed the
advanced arithmetic of the primary school in depth. In
the second year, arithmetic study continued with work
on ratio, proportions, rates, and powers and roots; two
hours a week were also devoted to the study of algebra.
This introduction to algebra included application of the
basic operations to integral expressions and work with
simple equations in one unknown. Mathematics study in
the third year was comprised of two hours' work each
week in algebra and geometry. Algebra study now included
solutions of systems of simultaneous linear equations,
work with fractional expressions, and simplification and
interpretation of quadratic expressions. Geometry
encompassed work with lines, the simple polygon, and the
circle. These studies were continued into the fourth
year with algebra, including irrational expressions,
ratio and proportion series (arithmetic and geometric
progressions), permutations and combination, Binominal
Theorem, and the use and theory of logarithms; geometry

[14]Ibid, pp. 232-234.

considered areas, proportion, and further properties of
the circle. The final year of secondary school also
required two hours of geometry study, but this time supplemented it with two hours of trigonometry. Properties
of planes, polyhedra, and solids with curved surfaces
were studied in geometry. The trigonometry course
included solutions of right-angled triangles, circular
functions, and use of formulae for the sum of angles.

The 1909 dual-track secondary program provided
additional mathematics instruction for science majors.
This additional study was in integral and differential
calculus.

Mathematics Learning Difficulties

The quality of mathematics instruction during this
period was relatively low. Edward Ross, observing
China's educational scene in 1911, described the reforms
as "all show, no results" and the teaching as an example
of "the blind leading the blind."[15] Teaching still
labored under the memory training techniques of the
traditional system. Class lessons fostered memorization
rather than developing independent thought processes.
An anecdote illustrating this phenomenon is given by Ross
in the story of a European teacher who introduced his

---

[15]Edward Alworth Ross, The Changing Chinese(New
York: The Century Company, 1911), p. 310.

mathematics class to logarithms. The following day, the students complained how difficult the previous night's homework was--they had tried to memorize the tables![16] A teacher noted that his students could learn proofs in geometry rather well, but "do not take quickly to mathematical reasoning."[17] At one period of time during this era of reform, the official mathematics curriculum required calculus to be taught in the junior year of high school. The literary chancellor of one province insisted that this regulation be followed at the local college, upon which the mathematics professor gave a few lectures on "the uses of calculus." Students were examined on the material and duly certified to be proficient in calculus.[18] While perplexed with the learning techniques of Chinese students, astute observers were not quick to condemn them, but rather took into account the conditions that nurtured such attitudes. One writer, in defense of the Chinese ability in mathematics, pointed out that the three top naval cadets in mathematics at Greenwich (England) were Chinese students.[19]

---

[16]_Ibid._, p. 337.

[17]_Ibid._

[18]_Ibid._, p. 338.

[19]_Ibid._, p. 334.

The Manchu Contributions to Mathematics Education

Although major educational reforms of the late nineteenth and early twentieth centuries stressed the need for mathematical studies, the school system of 1903-1912 did not respond accordingly. Curricula were still mainly humanistic. In the lower primary schools, eighteen hours a week of work were given on traditional studies as compared with six for mathematics; this ratio increased to twenty-two to three at the upper primary level. It remained high through secondary school-- fourteen to four. Of the foreign studies introduced, mathematics and English did occupy the most time. The distribution of mathematical subjects, arithmetic, algebra, geometry, and trigonometry, in middle school would remain basically the same for the next fifty years of Chinese educational history. Other characteristics remained the same for much of this time. These were an emphasis on the theoretical aspects of mathematics, a formalistic lecture approach to instruction, and a need-less repetition of many mathematical concepts in the school syllabus. Regardless of these difficulties, it was in this era that the concepts of mathematics educa-tion were put to work in China.

The Republican Period, 1912-1925

The wave of nationalistic fervor that ushered in
the Chinese Republic was reflected in a reorganization of
the school system. On January 9, 1912, a new Ministry of
Education was established in Nanking, and soon afterward
the official goals of education were revised. While
Chinese educational institutions were still strongly
influenced by the Japanese system, emphasis was now
given to manual labor and military training. Study of
the Chinese classics was eliminated from the primary
curriculum and the courses of study shortened: four
years of schooling was allowed for lower primary, three
years for higher primary, and four years for middle
school. Two-stream middle schooling was abolished. This
new system of education was designed to "produce strong
and well equipped citizens for the State." Lower primary
schools now became "Citizen's Schools" and their abbreviated curricula included moral studies, Chinese language,
arithmetic, manual arts, drawing, singing, and physical
training.[20] Higher primary studies added classics,
Chinese history, geography, natural science, and agriculture[21] to this program. Revision of the secondary curriculum was similarly undertaken. (See Table 4.)

---

[20]Teng and Lew, op. cit., Section III, p. 6.

[21]Girls studied housekeeping.

Immediate efforts to realize the new educational
policies resulted in the establishment of many more
schools and colleges throughout the country. Increased
political and civil strife and the lack of a strong
central government to implement the reforms combined to
decrease the efficiency of the scheme. Little progress
was made in the training of teachers or the improvement
of teaching methods. While "practical" was a key word
in the new plans, innovations such as practical science
teaching, handiwork, and physical training were adopted
in theory only, to satisfy government instructions. The
introduction of manual work into the school syllabus
received token support in the form of basket-weaving and
drawing courses. It was at this time that the intellectual climate was to be directed by a new force--returned
students from America.

TABLE 4

REPUBLICAN MIDDLE SCHOOL SYLLABUS
(HOURS/WEEK)[22]

Subject	Year			
	1	2	3	4
Ethics	1	1	1	1
Chinese	7	7	5	5
Foreign Language	7	8	8	8
History	2	2	2	2
Geography	2	2	2	2
Mathematics	5	5	5	4
Physics	–	–	4	–
Chemistry	–	–	–	4
Civics and Economics	–	–	–	2
Drawing	1	1	1	2
Handiwork	1	1	1	1
Music	1	1	1	1
Physical Training (military drill)	3	3	3	3

Beginning of American Influence in Chinese Mathematics Education

In the nineteenth century, American missionaries sent promising young Chinese to the United States for their higher education. This exchange was severely limited by the financial considerations involved in the undertaking. After the Boxer Rebellion, China agreed to pay an indemnity to the United States in reparation for damages incurred to American property and personnel during the

---

[22]Tan, op. cit., p. 129.

conflict. Rather than accepting these funds, the American government specified they be used to send Chinese
students to study in America. This decision removed the
financial impediment to an effective exchange and provided for an educational exodus. The United States
became the Mecca of enlightened modernism. Soon Chinese
students in America numbered in the thousands, but
unlike their predecessors who studied abroad in order to
master Western crafts, they sought "new learning" in all
disciplines, particularly education. Teachers College,
Columbia University, attracted many of China's future
teachers. At this time, the progressive education movement was being initiated at Teachers College by such men
as John Dewey and Paul Monroe. Profoundly influenced by
progressive ideals, Chinese students who came under their
influence returned to China and radically changed the
educational scene. Concepts such as "education for
democracy" and "student individuality" were now introduced into Chinese education.

Under the leadership of these returned students,
discipline-oriented education of the early Republican
period gave way to a concern for children's "freedom of
development." In general, however, the objectives of
mathematics education changed very little. Arithmetic
was still taught with the goals of training "quickness
of the mind and hand during exercises." Individual

teachers or schools experimented with such imported innovations as the Dalton Plan,[23] but these instances were
isolated and uncoordinated. The first large-scale American-inspired reform was introduced in 1917 by the Nanking
Teachers College. This college designed a new middle
school course of study which functioned on a semester
credit system. The program provided for four years of
work and allowed three-track streaming: arts, science, and
commercial studies. Nanking's program was approved by the
Ministry of Education. The complete syllabus as published
in its journal of April 1921 is given in Table 5.

TABLE 5

NANKING TEACHERS COLLEGE
MIDDLE SCHOOL PROGRAM[24]

Subject	Required Studies			
Year:	1	2	3	4
Semester:	1 2	3 4	5 6	7 8
Civics	1 1	1 1	—	—
Physical training	3 3	2 2	2 2	2 2
Chinese	7 7	6 6	3 3	3 3
English	7 7	6 6	2 2	2 2
Mathematics	6 6	6 6	2 2	2 2
Chinese History	3 3	—	—	—
Chinese Geography	—	3 3	—	—
Physiology and Hygiene	1 1	—	—	—
Nature Study	—	3 3	—	—
Drawing	1 1	—	—	—
Music	1 1	—	—	—

---

[23]A laboratory approach to learning based on student discovery. See John Dewey, The Dalton Laboratory
Plan (New York: Dutton, 1921).

[24]Peng Chun Chang, Education for Modernization in
China (New York: Teachers College Press, 1923), p. 9.

TABLE 5 (Continued)

Subject	1	2	3	4
Handiwork	–	1 1	–	–
Vocational Guidance	–	1 1	–	–
Chemistry	–	–	–	4 4
Physics	–	–	4 4	–

Elective Subjects

Arts Preparatory

Term:	1st	2nd
Chinese	5	5
English	3	3
Sociology and Economics	3	3
Elements of Logic	2	2
Elements of Philosophy	2	2

Industrial Preparatory

Term:	1st	2nd
Higher Algebra	3	3
Practical Mechanics	3	3
Manual Work	3	3
Mechanical Drawing	2	2
Engines and Motors	2	2
Materials Testing	3	–
Mechanical Engineering	–	3

A brief outline of the mathematical studies would in-clude:[25]

First two years (6 hour week)

Arithmetic: First term first year, three hours a week to supplement higher elementary school arithmetic, emphasizing fractions, percentage, ratio, and square roots.

Mathematics: First term, first year, three hours a week. Second term, six hours.

[25]Ibid., pp. 10-11.

Text: Breslick's Mathematics
      Volume I; second year,
      Breslick's Volume II.

Third and fourth years (2 hours week)

  Solid geometry: Third year
                Text: Wentworth and Smith

  Plane geometry: Fourth year
                Text: Wentworth

  Science or engineering elective for third and
  fourth years (3 hours week)

  Higher algebra: Third year
                Text: Hawkes

  Analytic Geometry: Fourth year
                   Text: Wentworth

This flexible but more specific curriculum was perhaps better designed to suit the needs of the changing Chinese society.

During this time of reformulation of Chinese educational philosophy, three Western educators of world renown, John Dewey, Paul Monroe, and Bertrand Russell, were invited to China by the Ministry of Education. The impact of these men upon Chinese educational thinking was decisive. Professor Dewey was known as a great apostle of philosophic liberalism and experimental methodology. An advocate of complete freedom of thought, he equated education with the solution of practical problems of civic cooperation and useful living. Dr. Monroe, with his unique knowledge and experience in teaching and school organization, was asked to examine the state of Chinese

schools. Russell was known as a man who had evolved a
philosophy of life and of the universe from science.
The philosophies of all three men stood in sharp contrast
with those of the conservatives who, even at this late
date, still exerted considerable influence in China.

Encouraged and reinforced by the appearance of
their mentors, American-trained educators concentrated
their drive for reform. A powerful lobbying group composed of many Dewey disciples from Teachers College was
founded, the National Association for the Advancement of
Education. In response to the pressures of this and
other groups urging reform, the National Federation
Conference of the Provincial Educational Associations,
held in Tsinan in September 1922, resulted in the formation of a new American-oriented school system. This
reform provided for four years of junior and two years
of senior primary schooling, six years of secondary, and
four to six years of university training. Secondary
school retained two of the tracks of the Nanking plan,
arts and science.

Standards for this new program were published by
the Commercial Press in 1925.[26] Of the time allowed for
the eleven required primary school subjects, 30 per cent

---

[26] Educational Association of China (E.A.C.), The
Outline Standards of the New System Curriculum (Shanghai:
Commercial Press, 1925).

was spent on national language, 20 per cent on social
studies, 10 per cent on arithmetic, and the remainder was
distributed among the other subjects.[27] Yǔ Tzǔ-i was the
mathematics specialist selected to design the outline for
primary school arithmetic. Yu's goals of arithmetic
teaching were to provide experience, accuracy, and speed
on behalf of the student in problem-solving. The scope
and sequence of his program included:[28]

First Year:

1. Solve problems in quantities by plan methods
   as opportunity offers; it is not necessary
   to use formal arithmetic.
2. Learn informally the terms in arithmetic,
   such as sum, remainder, how many, size,
   length, square, length, round, etc.
3. Write and read figures and symbols in
   arithmetic as opportunities offer.

Second Year:

1. Addition and subtraction with numbers below
   ten; adding numbers below ten to numbers
   below a hundred, filling numbers to a certain number in order to get ten, multiplying and dividing numbers by numbers in two
   figures; multiplication and division with
   numbers of two digits, computations of
   double and half a number, addition and subtraction with number of two figures. (No
   carrying in adding and no borrowing in subtracting.)
2. The ideas and terms of odd numbers, even
   numbers, length (such as Chinese feet and
   inches), quantities (such as Chinese
   piculs), square and cube, etc.

---

[27]Ibid., p. 4.

[28]Ibid., pp. 14-18.

3. The ways of writing and reading numerals
with two and three digits and forms of
addition, subtraction, multiplication, and
short division.

Third Year:

1. Same as second year, plus the addition and
subtraction of all kinds with numbers each
not exceeding ten; multiplying and dividing
tables of 2, 3, and 4; adding with carrying
and subtracting with borrowing; multiplication of tens with carrying and dividing
numbers which have remainders.
2. Ideas and terms of length, quantity, square,
cube, 1/2, 1/3, 1/4, etc.
3. Reading and writing numbers of 4 and 5
digits, and learning the Roman numerals.

Fourth Year:

1. Same as third year, plus multiplication
and division tables with numbers each not
exceeding 10; multiply numbers by numbers
of two digits and long division.
2. Terms and ideas: decimal, length, quantity,
square cube, time, money, 1/5, 1/6, 1/7,
1/8, 1/9, etc.
3. The ways of reading and writing of decimal
and denominate numerals.

Fifth Year:

1. Same as fourth year, plus the teaching of
exercises in the four operations, the
elementary denominate numbers, decimal
numbers, fractions, and percentage.
2. Understand the methods of reducing fractions; and the relations between fractions
and decimals; and between fractions, decimals, and percentage.

Sixth Year:

1. Same as fifth year, plus simple interest,
simple proportion, etc.
2. Understand the function of percentage, etc.

Yu's suggestions concerning the techniques of instruction
reveal his progressive orientation. Teachers were to
build upon informal play and social situations in
instilling an appreciation for the need and utility of
mathematics. Swiftness and accuracy in calculation were
to be derived from drill work. Problems which concerned
adult situations beyond the "imaginative ability" of the
students were not to be used. Emphasis was to be on
inductive rather than deductive reasoning.

Junior middle school mathematics was outlined by
Hu Ming-fu.[29] Elementary algebra and geometry supplemented by arithmetic and trigonometry comprised the
mathematical studies of the lower middle school. Hu
advocated the teaching of these subjects "correlatively
and equally," indicating a unified approach to mathematics instruction. However, his outline departmentalizes
the disciplines:

1. Arithmetic. The four operations, prime numbers, factors, divisors and multiples, H. C.
   F., L. C. M., fractions, decimals, ratios and
   proportions, square and square roots, area,
   volume, and interest.
2. Algebra. Symbols, terms and expressions,
   positives and negatives, four operations,
   linear equations, multiples and factors,
   fractions, simultaneous linear equations,
   quadratic equations, simultaneous quadratic
   equations, exponents, imaginary numbers, proportions, progressions, logarithms, and
   interests.

[29]Ibid., pp. 72-73.

3. Geometry. Axioms, theorems, straight lines, angles, perpendicular lines, parallel lines, triangles, parallelograms, polygons, circles, chords and tangents, constructions, areas, proportions, and similar figures.
4. Trigonometry. Measurements of angles, positive and negative angles, sines and cosines, tangents and cotangents, secants and cosecants, simple formulas, solutions with triangles, general ideas on the practical uses of trigonometry.

This mathematics was intended as a survey; it satisfied the needs of students who desired to follow an arts elective while providing a suitable foundation for those who chose a science major.

Proposals concerning the mathematics required of higher middle school students who elected a science option were written by Wang Kuei-jung, Ho Lu, and Nee Yao-shui. The branches of mathematics these gentlemen selected were trigonometry, geometry, algebra, and analytic geometry.

Trigonometry[30] was required for two periods a week for one semester, and carried a three-credit value. Wang advised that while the study should include practical problems, it should not stress surveying, as had been done in the past. Trigonometry's utilization in science and engineering was to be emphasized, and its relation with algebra and geometry made evident. References including Hobson, Granville, Wentworth and Smith, and

---

[30] _Ibid._, pp. 117-118.

Breslich and Stone[31] attested to an American influence.

The material to be covered included

The trigonometrical functions of acute triangles.
Solutions by right triangles.
The trigonometrical functions of any triangle.
The relations between the trigonometrical func-
   tions.
Oblique triangles.
Law of sine, cosine, and tangent.
The various properties of triangles.
Relations between the trigonometrical functions
   of the various angles.
Trigonometrical functions of the sum and differ-
   ence of two angles.
Trigonometrical functions of the double and half
   of an angle.
The inverse trigonometrical functions.
Trigonometrical equations.
The theory of infinity.
Exponential series and logarithmic series.
Method of compiling the logarithmic table.
The theory of complex numbers and the theorem of
   De Moiore.
Navigation.
Trigonometrical solutions for equations.

According to Ho Lu, the teaching of geometry was

to be flexible enough to include a consideration of non-

euclidean geometries, and provide an introduction to

analytic geometry through work on quadratic curves.[32]

He urged a logical development of the following list of

_____

[31]William E. Hobson, A Treatise on Plane Trigonom-
etry (Cambridge: Cambridge University Press, 1897);
William A. Granville, Plane Trigonometry and Four Place
Tables of Logarithms (New York: Ginn and Co., 1909);
Ernst R. Breslich and Charles Stone, Trigonometry with
Tables for Use in Senior High Schools and Junior Colleges
(University of Chicago Press, 1928); Trigonometry and
Tables (New York: Ginn and Co., 1915).

[32]E. A. C., op. cit., pp. 119-121

topics from definitions to theorems:

Plane geometry:

1. Point, line, angle, perpendicular and oblique lines, triangles, parallel lines, parallelogram, symmetric locus, construction.
2. Circle, chord and arc, central angle, inscribed angle of a circle, quadrilateral, relative positions of two circles, construction--perpendicular line, segmental line of an angle, parallel lines, tangent, common tangent of two circles; conditions for fixing a figure--methods of finding a line and a point, translation and rotation, transference of a plane figure, locus curve of a moving point.
3. Ratio and proportion: Characteristic of angular bisectors of a triangle, similar figures, similar triangles, homothetic figures, relations between the side of a triangle, definition of projection, definitions of sine and cosine, common trigonometric formula, finding the last proportional, finding the median proportional, finding the extremes, geometric solutions of quadratic equations, conditions for fixing a circle, finding a tangent circle, power of a point with respect to a circle, radical axis of two circles, radical center of three circles, cross ratio and harmonic ratio, transversal, pole and polar, inversion and its characteristic.
4. Regular polygon, method of computing pi, circumference, areas, polygons, and circles, polygons of equal areas.

Space:

1. Parallel lines and planes, right intersectioned lines and planes, bihedral angles, the common perpendicular of two straight lines, projections, trihedral angles, polyhedral angles, conditions of equal trihedral angles.
2. Polyhedrons--rhombus, cones, and their volumes, spacial symmetric figures, spacial similar figures.
3. Cylinders, cones, spheres, rotating bodies.

Quadratic curves:

> Ellipse, parabola, hyperbola, common properties
> of quadratic curves, construction.

Algebra instruction was to be centered around the concept of function, solutions and applications of graphical expressions, methods of study, and practical applications.[33] Concepts were to be given preference over the manipulation of symbols. Wang cautioned that the purpose of such directions was twofold: first, the training of students in fundamental thinking processes, and secondly, their preparation for higher mathematical studies. Algebra study consumed three periods a week for one year (6 credits). The rigor of this course was exemplified by a list of possible texts which included:

> Higher Algebra, Hawkes
> College Algebra, Fine
> College Algebra, Charles Smith
> Higher Algebra, Hall and Knight
> College Algebra, Wentworth
> Theory of Equations, Burnside

The topics to be studied were:

> Fundamental operations and principles.
> Factors.
> Fraction, highest common factor, lowest common
>   multiple.
> Partial fractions.
> Exponents and roots.
> Imaginary and mixed numbers.
> Logarithms.

---

[33]Drafted by Wang Kuei-jung, in *ibid*., pp. 122-123.

Ratio, proportion, and variation.
Permutations and combinations.
Theorem of quadratics.
Linear equations in one unknown.
Linear equations in two unknowns.
Determinants.
Functions and plotting graphs.
Quadratic equations in one unknown.
Fractional equations.
Irrational equations.
Inverse equation and binomial equation.
Quadratic equation in two unknowns.
Inequalities.
Infinitely great and infinitely small quantities.
Irregular equations.
Equation of logarithms and equation of exponents.
Theory of equations.
Equations of the third and fourth powers.
Arithmetical, geometrical, and harmonic progres-
    sions.
Infinite series.
Increasing and decreasing progressions.
Method of computing the sum of a progression.
Compound interest and annuities.
Important progressions, as exponents.
Logarithms, trigonometric expressions, etc.
Continued fractions.

The course in analytic geometry was designed by
Nee Yao-shui to be given in the final year of secondary
schooling.[34] It carried three credits, although it was
to be taught three periods a week for one year. Nee
made it clear that the course was to be of an introductory nature and that it would not go beyond work in two
dimensions. Solid analytic geometry would be studied
later in college. His brief outline included

Descartes' coordinate system and point.
Projections and their principles. Loci

[34]Ibid., pp. 124-125.

and equations, straight line and linear equations Ax + By + C = 0. The angle between
two straight lines intersected. The conditions for two straight lines being parallel
or perpendicular one to the other.
The straight line. Lines through the intersection of two given lines. Circle and quadratic equation.

$$Ax^2 + Ay^2 + 2Gx + 2Fy - C = 0$$

Polar coordinates. Transformation of coordinates. Conic sections and equations of the
second degree.

$$Ax^2 + 2HXy + By^2 + 2Gx + 2FY + C = 0$$

Polar formulas of conic sections. Tangent,
normal, second tangent, and second normal.
Poles and polars.
Natures of conic sections (including the
following three) Parabola, ellipse, hyperbola.
Higher plane curves.

The implementation of the new educational reforms

were difficult to realize due to the increasing social

upheaval then taking place in China. With minor varia-

tions, the patterns for the mathematics instruction of

several generations of Chinese students had been set.

Interestingly, no mention is made in this program of work

with the abacus; however, many schools still continued

this practice. Some middle schools, such as the Peking

Junior Middle School, offered the study of abacus

techniques as an elective, justifying this knowledge as

being needed in the commercial world.[35]

---

[35] Pae-Yeo Yuan, "A Comparative Study of Chinese
and American Secondary Education" (unpublished Master's
thesis, University of Pennsylvania, 1924), p. 40.

The Climate of Mathematic's Instruction

In conjunction with a progressive educational philosophy, American-trained educators brought to China an awareness of the necessity of testing and statistical analysis of educational reforms. Although standards for education existed on paper, no provisions had been made for determining their adherence. Responding to another invitation of the Chinese government, William McCall of Teachers College, Columbia University, founded a psychological laboratory at Nanking to construct tests for use in primary and secondary schools. An associate, E. L. Terman, working with L. C. Cha of the Peking National Normal University, devised a test based on Thorndike's New Psychology of Arithmetic to examine the mathematical reasoning processes of Chinese school children.[36] Sampling 92,000 students, Terman and Cha concluded that 40 per cent were studying mathematics at inappropriate levels. They concluded that 20 per cent of the students should have been demoted and 14 per cent promoted to a higher grade.[37] Influenced by this work, the Commercial Press published a series of standardized arithmetic tests. In a study to determine whether Western pedagogy

---

[36]E. L. Terman, The Efficiency of Elementary School in China (Shanghai: Commercial Press, 1924).

[37]Ibid., p. 165.

was suitable for the Chinese situation, Dr. Ai of the National Central University at Nanking employed the Curtis Arithmetic Test as a measuring instrument. The results of this testing demonstrated that the mental differences of the student population in question were not sufficient to warrant alternative teaching methods-- Chinese students could be taught in the same manner as American children![38]

While Dr. Ai relied on statistical affirmation of his theories, observers of the psychological and sociological climate of instruction questioned such policies. George Twiss, in his study of Science and Education in China,[39] noted that China's educational system was still bogged down in formalism. He observed that pupils were encouraged to be passive by the dictation of their teachers, who had no knowledge of teaching methods other than recitation. Many textbooks, especially in science and mathematics, were written in English. An example of one such work is the problem- and applications-oriented Middle School Arithmetic, by T. W. Chapman.[40] This book

---

[38]Morgan, op. cit., p. 47.

[39]George R. Twiss, Science and Education in China (Shanghai: The Commercial Press, 1925).

[40]T. W. Chapman, Middle School Arithmetic, Chung Hwa Book Co., Ltd., Shanghai, 1919. A similar format was found in Lo King Tuen and H. B. Graybill, Practical Arithmetic: A Mastery of Modern Mathematics for Chinese Middle Schools, Commercial Schools and Business Students, (Shanghai: Edward Evans and Sons, 1923).

begins by reviewing the four operations, with stress on
"mental exercises" and "short cuts." Decomposition of
number is employed in addition and multiplication. Subtraction is presented as complementary addition. Two
methods are given for division: factoring the divisor,
then successive division by the resulting factors, and
the "Italian method." Fractions, decimal fractions,
ratio, and proportion are treated in the usual manner.
Much of the text's contents concern commercial situations: brokerage problems, monetary exchange rates, etc.
One chapter is devoted to "Approximations and Contracted
Methods," and still another considers problems of time
and distance and time and work. An appendix contains
some 500 exercises taken from the British Civil Service
Examinations. A reader is left with the impression that
the text is designed to train clerks for the banking
houses of Canton or Shanghai, with mathematics becoming
a mental discipline, one requiring a well-trained memory.

Twiss offered his suggestions for curricula reform
and, in particular, mathematics teaching. He advised
that instruction should never be solely by lecture; a
laboratory method employing graphical techniques, geometrical constructions, and models was recommended. For
Twiss, the objectives of mathematics instruction were twofold:

1. To develop a skill in the use of mathematics as a tool, or instrument for economy
   in thought and calculation.
2. To acquire skill in logical thinking.[41]

He referred his Chinese colleagues to the books and
methods of the University of Chicago's Breslich, a
pioneer in the use of the mathematics laboratory.

Proponents of the "individualistic" approach to
education, based on an American model, assumed that a
competitive climate of intellectual activity could be
generated among Chinese students. Such an attitude was
basic to a progressive philosophy of education, but
difficult to stimulate in China. Chinese teachers and
students were deeply concerned with the image they
projected to their fellow man. Westerners have labeled
this phenomenon "face." Thus to preserve "face," a
teacher never admitted a lack of knowledge concerning a
student's question. As a precaution against such challenges, questions were discouraged. Similarly, students
would not risk losing "face" by venturing opinions or
queries that would make them appear ignorant. This
preservation of a social image and maintenance of the
resulting status quo contradicted and defeated many of
the proposed aims of the "New System."

----

[41]Twiss, op. cit., p. 266.

Opposition to the "New System"

Conservatives saw these reforms as a clear break
with traditional Chinese educational principles. Dewey's
experimentalism and Russell's "materialistic immoralism"
were viewed as diametrically opposed to China's cultural
heritage.[42] Even as late as the second decade of the
twentieth century, two of the most progressive centers
in China, Nanking and Canton, still had large numbers of
Confucian-oriented schools.

More objective critics based their opposition to
popular educational practices on China's evolving social
needs. For them China was a developing nation and as
such required an educational system designed to produce
knowledgable technicians, rather than university candidates steeped in theoretical knowledge. In a work
published in 1923, Education for Modernization in China,
a Columbia University alumnus, Pêng-chun Chang, offered
an alternative plan. Chang wanted education to produce
leaders for the new state by training students in
exploring and pioneering, new community building, and
scientific production and organization. In achieving
these goals, he wanted to utilize teaching methods
representative of Dewey's experimentalism, such as

---

[42]Kao Chien, "Progressive Education Undermined
China," The Freeman (December, 1954), pp. 216-218.

employed in the Dalton Plan or the Interlaken School[43]

scheme. His students were to learn vocational skills by

actual participation in factory and agricultural activi-

ties. Instead of listing specific subjects to be studied,

Chang provided a broad outline of suggestions as to how

certain disciplines might be directed towards his goals.

Mathematical studies were grouped under three headings:

> Exploring and Pioneering. Practicing arithmet-
> ical calculations in connection with the needs
> of providing food, shelter, and clothing;
> learning to keep simple accounts for individual
> use and for the school community; practicing
> surveying, learning elements of geometry and
> trigonometry for surveying and map-making;
> making plans for the foundations and structure
> of some building, learning geometrical drawing
> in the design; learning simple methods of pre-
> senting facts on charts and graphs in the
> process of making reports.
>
> New Community Building. Learning mathematical
> symbols and functions in connection with the
> planning and construction of city streets, city
> sewage, city lighting, city telephoning and the
> like: practicing keeping community accounts
> including budget making, bookkeeping, and
> auditing; using graphs and charts in propaganda
> material and reports.
>
> Scientific Producing and Organizing. Learning
> mathematical tools in connection with the needs
> of the different kinds of practical work in the
> productive vocations, according to individual
> interests and aptitudes.[44]

---

[43]A school that incorporated vocational skill with
academic learning. See John Dewey and Evelyn Dewey,
School of Tomorrow (New York: E. P. Dutton and Co.,
1915), pp. 87-89.

[44]Chang, op. cit., pp. 75-79.

Unheeded in the reforms of this period, a variant of Chang's program would become popular with a future government of China.

The Republican Period in Retrospect

As with the preceding periods of educational reform, written words exceeded concrete manifestations. A survey taken of forty-seven middle schools revealed that at the lower level only 56 per cent were offering algebra, 38 per cent composite arithmetic, 27 per cent trigonometry, and 19 per cent arithmetic. For upper middle schools the statistics were about the same: algebra 69 per cent, geometry 65 per cent, composite mathematics 41 per cent, arithmetic 38 per cent, and analytic geometry 25 per cent.[45] Even in normal schools, only about 7 per cent of instructional time was spent on mathematics.[46]

The reforms of the time echoed the newly acquired spirit of Western liberalism. Returning students from overseas brought with them the ideas of building a new China, and in the process attempted to destroy the old.

----

[45] Earl H. Gressy and C. C. Chih, East China Studies in Education, No. 5: Middle School Standards (Shanghai: East China Christian Education Association, 1929), p. 33.

[46] "Required Subjects of Six Lower Normal Schools in China," The Educational Magazine (Shanghai, May, 1922), pp. 1-43.

Nationalism and its resulting anti-foreignism was perhaps the most profound movement of the time. Disillusionment with the image of Western society and institutions as presented by World War I and its resulting treaties prompted many young Chinese to turn from the West as a source of enlightenment. Some turned back to the traditional learning of China. Still others sought direction in the writings of Marx and Lenin; Marxist study groups were formed and the Communist party was founded in China. The period was one of political, social, and economic unrest. In the vacuum caused by the collapse of the monarchy, political and military opportunists strove to enhance their positions. As a state and as a society, China became fragmented. Warlords controlled many provinces. Rural areas became alienated from urban centers as never before, due to the industrial and intellectual revolutions taking place. Returned students clustered in the few areas of China that exhibited foreign influence--the large cities, such as Shanghai and Canton. These cities became the focal points of the new learning. If the establishment of modern schools and new reforms were realized, it was in these few cities. Instead of uniting the country, the new education did much to divide it. A more complete institution of the Republican educational reforms would have to wait until a stronger degree of unification was achieved in the

country. This unification would be briefly realized
under the Nationalist government.

The Kuomintang Era: 1925-1949

As the number of young Chinese educated in the
modern schools increased, their desire to establish a
national identity grew. Gradually a political party
developed which established as its goals national unification and full independence from foreign influence.
This party was the Kuomintang, founded by Dr. Sun Yat-
sen. Its credo became Sun's San-min Chu-i, Three Principles of the People: People's Nationhood, Democracy,
and Livelihood. By the time of Sun Yat-sen's death in
1925, the Kuomintang had developed into a strong political and military machine. Its armies, under the direction of Sun's lieutenant, Chiang Kai-shek, inaugurated a
series of military campaigns that resulted in the defeat
of the warlords and assisted in reducing regional fac-
tionalism. Chiang established a nationalist government
in Nanking in April of 1927. From this new capital, he
directed widespread attacks against forces still outside
the government's sphere of influence. By 1930, a degree
of national unity had been achieved that was unique in
China's twentieth-century history. A contributing factor
in this unity was the Kuomintang's uncompromising use of
education in promoting the principles of the San-min Chu-i

A Series of Curricular Reforms

The Second Convention of the Nationalist (K.M.T.)
Party in January 1926 emphasized the party's concern with
education:

> There must be emphasis on education and it must
> be made revolutionary and proletarian, with
> special emphasis on the extension of mass-educa-
> tion, whilst the nationalist administrators must
> also make positive efforts to recover authority
> in education.[47]

This authority was sought through a highly centralized
system of educational administration. To bring order out
of chaos, the government prescribed new standards of
curricula, equipment, finance, and examinations to which
all schools were expected to conform. Education was
directed at promoting the San-min Chu-i.

By September of 1928, the Kuomintang revised the
educational system. The new school curricula differed
little from Republican models. All levels of education
were to attain a close relationship with local industry,
where students were to obtain practical industrial
experience. The main objective of middle schooling was
officially declared to be university preparation.
Intensified study of mathematics and the English language
were urged to accomplish this goal. While the study of
English and mathematics was increased, the proposed
factory experience did not materialize.

---

[47]Peake, op. cit., p. 93.

In September of 1931, at the request of the
Nationalist government, the League of Nations sent a
mission to China to assist in educational reconstruction.[48] After three-months' investigation, the mission
published its report.[49] Included in its findings was a
strong criticism of China's "elitist" school system,
which was based on the imitation of foreign institutions
quite alien to Chinese culture and the masses of Chinese
people. Curricula were described as "overburdening" and
teaching styles as formal and "bookish." Secondary education was providing inordinate numbers of journalists
and rhetoricians whose employment would be questionable.
Professor Tawney, one of the League's observers, suggested that the primary curriculum should be restructured
around village life, agriculture, and practical work.[50]

The Ministry of Education revised the primary
school program in March of 1932 and the secondary program
a year later. Both programs strove for more rigor by

---

[48]The members of this mission were: Prof. Carl H.
Becker, University of Berlin; Prof. M. Falski, Director
of Primary Education, Poland; Prof. P. Langevin, College
de France; and Prof. R. H. Tawney, London School of
Economics and Political Science.

[49]C. H. Becker, The Reorganization of Education in
China (Paris: League of Nations, Institute of Intellectual Cooperation, 1932).

[50]R. H. Tawney, Land and Labour in China (London:
George Allen and Unwin, Ltd., Museum Street, 1932), p.
187.

increasing subject hours sharply. An attempt was made to
insure school standards by instituting graduation examinations at the junior and senior levels of middle school.
The results of the examinations were disappointing, and
the government was forced to restructure its educational
system in October, 1935. Now some of the League's suggestions were heeded. The Conference for Revision of
School Curricula recommended a reduction in subject
hours for middle schools, an increased emphasis on
patriotism, and a return to a dual-track middle school.
Streaming in middle school had been abolished in the
1933 plan. This plan was also short-lived as Japanese
aggression forced China into wartime conditions. By
1940, the Ministry of Education had devised a program
of wartime education that stressed political indoctrination, physical exercise, and military and paramilitary
training. For economy, streaming at the middle school
level was eliminated--all students studied the same subjects. For the distribution of subject hours in various
Kuomintang middle school programs, see Table 6 below.

TABLE 6

THE DISTRIBUTION OF MIDDLE SCHOOL SUBJECT
HOURS FOR KUOMINTANG EDUCATION PROGRAMS

Subjects Program:	1928	1933	1935*		1940	1947
Civics	18	22	14		14	10
National Language	60	66	64	(67)	64	60
Foreign Language (English)	46	60	64	(67)	60	50
History	24	26	24		20	20
Geography	18	12	24		16	18
Mathematics	49	48	56	(48)	52	48
Nature Study	15	18	20		16	6
Hygiene	4	8	2		4	4
Drawing	6	20	9		18	14
Music	6	14	9		16	16
Physical Education and Scouting	18	30	36		36	30
Industrial Arts	9	16	12		16	16
Chemistry	8	13	12		12	10
Biology	6	10	8		–	16
Physics	8	12	12		12	16
Military Training	9	12	6		18	–
Electives	18	–	–		–	24
Logic	–	2	–		–	–

* Hours in brackets represent required hours for arts major.

The destruction of facilities and the privations of war seriously hampered the accomplishments of the wartime schools. Classes became overcrowded, textbooks were lacking, and the teacher shortage became even more acute as many teachers left the schools for lucrative government positions. To compensate for this breakdown of formal education, a variety of part-time and People's

Schools were established. These schools were especially effective at the primary levels where abbreviated instruction concentrated on "general knowledge" and mathematics.

After the completion of hostilities in 1945, the Ministry of Education considered the adoption of a five-year plan for educational reconstruction. A revised version was accepted in December 1947. Included in this plan were provisions for full-day primary and middle schools. This plan was in effect until the Nationalist withdrawal from the mainland.

Mathematics Instruction During the Period 1925-1949

While the Ministry inaugurated series after series of curricula reforms specifying course content and duration, it appears that individual schools' adherence to official standards varied. An analysis of the mathematics offerings of several middle schools of this era reveals that a school with high academic standards would have had a mathematics program of the following nature:[51]

Junior Middle School

General Mathematics. A unified course consisting of the fundamentals of arithmetic, algebra, geometry and numerical trigonometry. This study included:

---

[51]Adapted from University of Nanking Middle School curriculum for 1931.

Linear equations in one unknown.
Ratio and proportion.
Quadratic equations in one unknown.
Numerical trigonometry, use of sine, cosine,
   and tangent.
Rational use of significant figures.
Graphic representation of statistical data.
The linear function, y = mx + b.
Graphic solution of problems.
Empirical curves, fitting to observations.
Simple formulae, meaning and use.
Negative numbers, meaning and use.
Simultaneous linear equations in two unknowns.
Graphs as methods of representing dependence.
Computation with approximate date.
Variation.
Use of tables other than logarithms.

Suggested texts: New System Series of Correlated
   Mathematics for Junior Middle Schools, Commercial
   Press;

   Junior Middle School Arithmetic, Chung Hwa Book
   Co.;

   Mathematics, Breslich, Book I and II.

Advanced Middle School Mathematics. A continuation
of the combined approach to mathematics instruction covering work in algebra, plane and solid
geometry, and plane trigonometry.

Senior Middle School

Combined Mathematics. Review of the general and
advanced mathematics of the junior middle school.
Stress upon computation employing logarithms.

Algebra. Formal course at the level of Went-
worth's Algebra, topics covered included:
  Basic algebraic operations.
  Special methods in algebra.
  Factors, highest common factor, lowest common
     multiple.
  Simple and fractional equations.
  Simultaneous simple equations.
  Inequalities.
  Involution and evolution.
  Theory of indices.

Radicals, radical equations, surds.
Quadratic equations.
Simultaneous quadratic equations.
Roots.
Quadratic equations graphic representation.

Under the two stream system, an engineering
or science major would also take another
algebra course at the level of Davisson's
College Algebra.

Plane and Solid Geometry. A two course sequence
based on the contents of a book such as Wentworth's Plane and Solid Geometry considered:

Plane	Solid
The straight line	Lines and planes in space
The circle	Polyhedron
Proportion	Cylinder
Similarity and con-	Cone
gruence of polygons	Pyramid and sphere
Area of polygons	
Rational polygons	

Trigonometry. This course included:

Functions of acute angles.
Properties of right triangles.
Methods of angle measurement.
Problems of measurement.
Construction and use of tables.

Suggested text: Rothrock's Elements of Plane
and Spherical Trigonometry.

When a system of electives existed, a choice such
as the following might be offered:

Analytic Geometry, at level of Love's Analytic
Geometry.
Abacus Practice
Hard Question in Mathematics (Review) based on
a book such as Kong Honh-tah's Three Hundred
Hard Questions in Mathematics.
Elementary Functions and Their Applications,
based on a book of same name by Gale and
Watkey.

These courses, reflecting a strong American influence, satisfy the initial 1922 reforms. It appears there

was much unnecessary repetition at the lower levels of middle school, while upper division courses were perhaps ambitious in their conception. If such programs thrived, it could only have been in a few urban centers in the years of relative peace and unification, 1931-1935.

A New Facet of Mathematics Education: Political Indoctrination

May 15, 1928, saw the First National Education Conference at Nanking convened by Ts'ai Yüan-pei, President of the University Council. At this meeting, a motion was passed asserting that all future teaching material would be selected so as to emphasize the humiliations to which China had been subjected in its international relations of the preceding half century.[52] This action on the part of the conference initiated a new phase in Chinese education, political indoctrination. Whereas, moral indoctrination promoted the cause of society in traditional China, Nationalist China would employ political instruction to strengthen the state.

Revised curricula contained the doctrines of the Three Principles of the People and fostered an antiforeign bias. Even the instruction of arithmetic and drawing were to be viewed in a patriotic perspective.

---

[52]Cyrus Peake, Nationalism and Education in Modern China (New York: Columbia University Press, 1932), p. 4.

Close governmental censorship of the Commercial Press's
Hsin Shih-tai (New Age Series) assured this policy. An
example of politically oriented mathematics material is
given by The Popular Development National Language
Readers in which reading and arithmetic are combined to
provide a lesson in national consciousness:

> Two important questions of arithmetic should
> be met by any citizen of China:
> 1. If a nation being defeated in a war should
> indemnify 450,000,000 taels to the winning
> party of eight nations and make annual pay-
> ments evenly, over a period of thirty-nine
> years, how much would be the amount paid yearly?
> And how much would be the sum received by each
> of the victorious if the sum were divided
> evenly?
> 2. If there is a house entered by some brigand
> every year, and looted of Mex. $1,200,000,000
> and nothing is provided to prevent them, the
> loss may naturally increase annually, and in
> ten years it may increase by two and a half
> times. How much will then be the sum thus rob-
> bed from this house?
> Four hundred million people reside in this
> house, but only one out of ten earns money
> since there are among them women and children
> and the aged. If the money thus robbed were
> earned by these men, how much would be the
> annual sum that they should earn for the
> brigands?[53]

A Reappraisal of Educational Goals

In the beginning of the Nationalist era, the
objectives of education were highly idealized, with mathematics instruction directed toward preparation for higher

---

[53]The Sokokusha, Anti-Foreign Teaching in New
Textbooks of China (Tokyo: 1929), pp. 129-130.

studies. By 1949, such ideals had mellowed under years
of war to the realization that the mathematical knowledge
taught in school had to be basic, popular, and practical.
Education became less and less of an elitist activity.
Throughout the war years, while many school subjects
were eliminated or drastically reduced in study time,
mathematics remained a basic subject. In 1947, the
Ministry of Education suggested that the objectives of
mathematics instruction should be to provide China's
young with the mathematical skill they needed to make a
living. A utilitarian approach to mathematics teaching
at the secondary level was urged. Geometry was to be
related to construction plans. Middle school algebra
and analytic geometry were simplified, and in the final
semester of senior middle school a review of all mathematics studied was instituted.[54]

Conclusions

The first half of the twentieth century witnessed
a period of great educational flux in China. A modern
school system had been established, but its structure and
content were foreign imports: Japanese and American, and
thus did not necessarily reflect the needs of the new

---

[54]China Handbook 1952-53 (Taipei, Taiwan: China
Publishing Co., 1954), p. 233.

Chinese nation nor its people. Successive governments
from the Manchu through the Republican to the Kuomintang
revised their educational programs to satisfy constantly
changing national goals; however, internal social and
political fragmentation provided an unstable climate for
meaningful reforms. Despite the urgent quest for modernity and the affirmation of progressive ideals, traditional
educational thinking still predominated the school scene.

Education was elitest and examination oriented.
Curricula were formal and overburdening. Specifically,
mathematical studies bore little relevance to the requirements of an agrarian society or industrial application.
Methods of instruction encouraged rote memorization and
provided little opportunity for individual inquiry. A
student's accumulation of mathematical knowledge, while
facilitating his ascension to a higher education, did
little to acquaint him with the realities of his world.
Thus just as the spirit of Confucian teaching became lost
in stylized pedantry, so now the quest for scientific
enlightenment became smothered by formalism.

CHAPTER III

THE POSITION OF MATHEMATICS IN THE ESTABLISHMENT
OF A SOCIALIST SYSTEM OF EDUCATION

Yenan: The Formative Stage of Educational
Experimentation, 1935-1949

Launched as a political entity in 1921, the Communist Party soon became a prominent revolutionary force,
disciplined and highly organized. Peasants burdened with
extortionate rents and other landlord exactions welcomed
Communist land reform schemes. In 1926, the Communists
cooperated with the Nationalists in the Northern Expedi-
tion against the warlords, but by the following year,
they had split with the government and asserted their
independence. Under the leadership of Mao Tsê-tung, a
"Soviet State of China" was established in the Hunan,
Kiangsi, Fukien border region.[1] Socialist experiments
patterned on Russian experience were attempted in collective farming schemes and youth training programs.
General schooling was similar to that given in government
areas, with the exception of new educational goals.
Education would now be:

---

[1]Mao's decision to occupy border regions was based
on the premise that these areas, long disputed by neigh-
boring provinces lacked political and military integrity,
they were in principle "no-man's lands" awaiting political occupancy.

1. Directed towards the training of military personnel.
2. For the development of natural resources.
3. Designed to promote industrialization.
4. Applicable to agrarian reform. [2]
5. Promotion of a Soviet ideology. [2]

Routed from their Hunan-Kiangsi stronghold by K. M. T. armies, the Communists were forced to retreat northward in 1934. The resulting "Long March" terminated in Shensi a year later. Here Mao founded his new Communist state, with a capital in Yenan. Due to its geographical isolation, the region had long been removed from government influence. This resulting neglect allowed ready inroads by Communist ideology. In this physically hostile environment, isolated and surrounded by enemy forces, the basic tenets of Communist educational theory would be established. Mao Tse-tung, himself trained as a teacher, [3] would provide the motivating force in the formulation of an educational philosophy.

Mao on Education

Mao's epistemological theories contained in On Practice, [4] written in July of 1937, were reflected in the

---

[2] Chen An Ren, Modern Political History (Shanghai: Chung Hwa Book Co., 1943), pp. 217-219.

[3] Mao attended the First Normal School of Hunan, 1912-1917.

[4] See Anne Fremantle, Mao Tse-tung; An Anthology of His Writings (New York: Mentor, 1962), p. 200.

new "border area" education. Influenced by Marxist-
Lenist dialectical materialism, Mao proposes that all
knowledge is obtained through productive activity. Man's
collective class struggle results in production and
promotes revolution. Mao divides the cognitive process
into two main stages: the perceptual and the conceptual.
Through practice man perceives certain phenomena, but
his initial perceptions are not profound and cannot
result in meaningful conclusions. Further practice
leads to repeated perceptual exposure and produces a
qualitative leap in knowledge, giving rise to conceptualization. Concepts are formed and the essence of
things understood. Upon these concepts, methods of
judgment and logic can operate, producing a rational
knowledge in which a "wholeness" of essence is understood. Perception and reason, while being distinct
processes, are united on the basis of practice. Rational
knowledge obtained by practice is then directed back
through practice in the form of production. Production
will bring about a change or revolution. This cycle of
cognitive activity is repeated, with each cycle resulting
in a higher form of intellectual and productive accomplishment. Education's involvement in the acquisition of
knowledge must be initiated in objective reality through
practice, and concluded in the application of that
experience, production.

Mao's personal concern with education is evidenced by his numerous and varied statements on the subject.[5] In the translation of his philosophy to a workable pedagogy, he advocates a proletarian relevance, in contrast to the elitist nature of traditional Chinese educational practice. An avoidance of formalism is paramount. Concerning teaching methods, Mao cautions teachers to:

1. Resort to the method of enlightenment.
2. [Teach] from the short range to the long range.
3. [Teach] from the superficial to the deep.
4. Speak in the popular language.
5. Be explicit.
6. Make what you say interesting.
7. Aid speech with gesticulations.
8. Review the concepts taught last time.
9. Employ a teaching outline.
10. Adopt the method of discussion in a study course for cadre.[6]

Education had to be tempered with politics. The political aspect of education was viewed as a "central link" binding the student to the state. Mao considered it inadvisable to have too many subjects, but urged that class education, party education, and work had to be

---

[5]"Chairman Mao on Revolution in Education" (Shanghai: People's Publishing House, 1967); translated in Current Background No. 888, American Consulate General, Hong Kong (August 22, 1969).

[6]Mao Tsê-tung, "Resolution of the 9th Congress of the 4th Army of the Red Army of the Communist Party of China," December 1929; Current Background No. 888, American Consulate General, Hong Kong (August 22, 1969), p. 2.

greatly strengthened. Under this theory, educated youth

had to be taught to

1. Master Marxism-Leninism and overcome bourgeois
    or petty-bourgeois thought.
2. Acquire a sense of discipline and organiza-
    tion and oppose anarchism and liberalism in
    organization.
3. Make up their minds to go deep into the grass-
    roots level to carry out work and oppose the
    tendency to slight practical work experience.
4. Contact the workers and peasants, serve the
    workers and peasants resolutely, and oppose
    the consciousness [temptation] to despise the
    workers and peasants.[7]

These ideas were to be a molding force in the educational

policies of the Yenan and other border regions.

Establishment of Educational Facilities

The Nationalist government had little educational

success in Yenan. Before 1937, there were only 120

schools in the border regions.[8] Education was viewed

with suspicion by the natives and thought to promote a

disdain for labor. In 1937, a newly formed Communist

Education Department for the border regions issued guide-

lines for the establishment of elementary schools. A

five-year program, three years lower, two years upper,

[7]Mao Tsê-tung, "Instruction of the Military Affairs
Committee of the C. C. P. Central Committee on the Question of Consolidating the Anti-Japanese Military and
Political College," July 1939, Current Background No.
888, American Consulate General, Hong Kong (August 22,
1969), p. 4.

[8]Peter J. Seybolt, "Yenan Education and the Chinese
Revolution" (unpublished doctoral dissertation, Harvard
University, Cambridge, Mass., 1969), p. 231.

was agreed upon and model schools set up. Although
these schools functioned with a minimum of equipment,
they did provide weapons for drill, indicating the
increased emphasis on military activities. War was now a
fact of life in China. Improvisation and ingenuity were
fundamental to this new scheme, attested to by a statement by Hsu T'e-li, C. C. P. Education Commissioner:

> When there is no chalk, use charcoal or mud as
> chalk; when there are no writing brushes or
> paper, practice writing on the sand or dirt;
> when there are no books rely on your power of
> comprehension and your memory. As long as the
> teacher has one book, education can proceed.
> Benches and tables are even less of a problem,
> and when there is no school room teaching can
> be carried on out of doors.[9]

Lack of textbooks presented an especially critical problem. Teachers were encouraged to write their own materials based on the students' needs and experience.
Mao's policy of "proceeding from near to far, from
simple to complex, from concrete to abstract" was adhered
to whenever possible. Formal education was divided into
three categories--political, military, and cultural.
Included in cultural courses were common knowledge of
history and geography and common knowledge of nature and
arithmetic. A typical program of study for a Yenan lower
school was given in a July 1938 issue of the New China
Daily News. The six subjects listed were Chinese

---

[9]Ibid., p. 236.

language, arithmetic, political common knowledge, military

affairs, common knowledge of history and geography, and

singing and art.[10] The newspaper article went on to

criticize the teaching of arithmetic as not supplying

practical accounting skills, ones that might be immedi-

ately useful to the students' families. By 1942, the

government of the border area was more firmly established

and issued a standardized curriculum. The duration of

primary schooling was again returned to six years. See

Table 7 for the specific subjects studied in this program.

TABLE 7

YENAN PRIMARY CURRICULUM
1942 (HOURS/WEEK)[11]

Subjects	Year					
	1	2	3	4	5	6
Border Region Construction	4	4	4	–	–	–
National Literature	5	5	5	5	4	4
Mathematics	4	4	4	4	3	3
History and Geography	3	3	3	3	–	–
Nature Study	3	3	3	3	–	–
Political Common Knowledge	–	–	–	3	3	3
Production Knowledge	–	–	–	–	3	3
Medical Knowledge	–	–	–	–	3	3
Total Hours	19	19	19	18	16	16

Border Region Construction was a civics course

stressing service to the workers and peasants. Production

10Ibid., p. 240.

11Ibid., p. 305.

knowledge may have consisted of tending the school

vegetable garden or manufacturing a small item in the

school workshop. Resulting products could be sold, con-

tributing to the schools' income and promoting self-

sufficiency.

Subject matter in the lower middle school curricu-

lum was similar to that studied in K. M. T. regions. (See

Table 8)

TABLE 8

YENAN LOWER MIDDLE SCHOOL CURRICULUM
1942 (HOURS/WEEK)[12]

Subject	Years: 1	2	3	Total
	Terms:1 2	3 4	5 6	Hours
Civics	2 2	2 2	2 2	12
National Language	6 6	6 6	6 6	36
Foreign Language	4 4	4 4	4 4	24
Mathematics	4 4	5 5	5 5	28
History	2 2	2 2	2 2	12
Geography	2 2	2 2	2 2	12
Zoology	2 2	— —	— —	4
Botany	2 2	— —	— —	4
Physics	— —	3 3	— —	6
Chemistry	— —	— —	3 3	6
Physiology and Hygiene	1 1	1 1	1 1	6
Art	1 1	1 1	1 1	6
Music	2 2	2 2	2 2	12
Military Training	2 2	2 2	2 2	12
Total class hours	30 30	30 30	30 30	180
Independent study hour	17 17	17 17	17 17	102
Total School Hours	47 47	47 47	47 47	282

A striking feature of this program is the total number

of hours involved, 282. A comparable contemporary K.M.T.

---

[12]Ibid., p. 292.

program contained 246 hours of work.[13] This is under-
standable when it is realized that this was a terminal
program for most students in the border region.

Activities in education now spread from the main
Communist stronghold in Yenan to the "Liberated Areas."
This new territory was comprised of border regions
evacuated by the retreating Japanese and seized by Communist forces. It included the Shensi-Kansu-Ninghsia
area, the Shensi-Chahai-Hopei area, the Kiangeu-Anhswi-
Chekiang area, the Shensi-Hopei-Shantung area, and the
Northeast China area. In these regions the main thrust
of educational effort was not directed at general education. Priority was given to cadre training, adult education, and the education of children respectively.
Experiments at all three levels helped produce institutions and practices that would become fundamental in
later Communist educational methodology.

Chêng-fêng Sessions

In the preparation of courses for the training of
political cadre, teachers and staff members came together
in a "chêng-fêng" session. Such meetings took the form
of a struggle, within which the content and methods of
instruction were carefully scrutinized, criticized, and

---

[13]See page 98.

re-examined until no questions were left to chance.
Under the psychological pressures of group dynamics,
flaws and weaknesses in teacher presentation were
eliminated. Chêng-fêng struggles took the form of
criticism and, more importantly, self-criticism. While
these micro-purges tested political convictions, they
also served to reduce the tendency of face saving among
teachers. Thus the way was prepared for more informal
teaching situations. Without the teacher's need to
preserve a self-image, questions might be more freely
asked in the classroom. Classroom work became a collective rather than an individual endeavor. Examination
questions were discussed with students and study reference agreed upon before testing. This procedure was
designed to discourage "championism" among the students.
Chêng-fêng refinement was perfected at the cadre level
of education, and eventually employed in secondary and
elementary schools. As a result, teachers now began to
stress group leadership over individual dictatorship and
substituted "explanation and persuasion for beating and
cursing." Students were divided into study groups, with
each group having its own "life-direction" to follow.
Even with a new outlook on teaching, most teachers were
poorly qualified for their positions. The Party's practice of selecting the more qualified of its youth for
positions other than teaching did little to remedy the
situation.

116

Resolving the Needs for Relevance and
Popular Appeal

Results obtained from employing a standardized curriculum were disappointing. Objections to the Communist schools were analogous to those encountered by the K.M.T. system. Illiteracy was still rampant. Formalism prevailed in instruction. An article in the Liberation Daily of May 1944 was specifically concerned with the conditions of language and mathematics training:

> For example, Chinese and mathematics have traditionally been considered the most important courses in the school curriculum. But the middle school students who have studied them for five or six years can still not properly undertake the job of a reporter for a wall newspaper or of an accountant for a cooperative. This is because the line of education in the past was to train child literati or child mathematicians and not to train people who understood the subject and could work in practical life at publicity or computation.[14]

In response to such criticisms, the school system was extended to include min-pan or "people's-schools." Mao directed that "in our education we must have not only regular primary and secondary schools, but also scattered, irregular village schools; newspaper reading groups, and literacy classes."[15] Following the slogan "develop production, expand the schools," these part-time

---

[14]Seybolt, op. cit., p. 295.

[15]Mao Tsê-tung, "The United Front in Cultural Work," October 20, 1944, Current Background, op. cit., p. 6.

village schools were established under the direction of
local cadre. Some met for half-day (8:00 a.m.-12:30
p.m.) sessions, while others held classes on alternate
days of the week, allowing a traveling teacher to serve
two schools.[16] A six-day school week was followed. The
curriculum was designed to be "living," simple, pragmatic, relevant to the needs of the people, and within
the competence of the teacher. The two courses most
stressed out of a possible three or four were national
language and arithmetic. In the establishment of one
such school, the teacher consulted with the students'
parents, asking them what knowledge they believed their
children should have.[17] Predominate in their responses
was contract and letter writing and calculation of
accounts. Responding to these requests, the school curriculum was designed consisting of studies in the Chinese
language, mathematics, and health practices.

Arithmetic, being a priority subject, was usually
taught first in the morning. Work began with the learning of number names and the writing of numerals and the
names and application of various systems of Chinese

---

[16]Michael Lindsay, Notes on Educational Problems
in Communist China, 1941-47 (New York: Institute of
Pacific Relations, 1950), p. 85.

[17]Seybolt, op. cit., p. 269.

weights and measures. Computation, both with pencil and
paper and abacus, was employed. Practice was undertaken
on problems concerning grain measurement, taxes, family
budgets, and commercial accounts. Knowledge thus
obtained was applied by requiring the students to do
their family accounts and by sending some of them to
assist in the accounting work of labor brigades, weaving
cooperatives, and other organizations from which they
could gain practical experience.

Mathematical Content in the Curricula

While this period of Chinese Communist educational
history was extremely turbulent, some impressions can be
obtained from surviving texts concerning the level and
nature of primary mathematics instruction. Methods and
duration of teaching varied from area to area according
to local conditions. Each area had an educational committee which attempted to coordinate teaching activities
and provide for instructional materials. One such committee in the Shensi-Kansu-Liangshung region published a
series of six texts for primary arithmetic.[18] The
existence of six separate books conveys the impression

---

[18]Shensi-Kansu-Liangshung Border Area Education
Department, Primary Arithmetic, 6 vols. (Shen-fu Book
Store, 1946).

that their contents represented a complete primary school mathematics program. Instructions, however, given in the preface of the last volume, state that each book is intended to provide material for an eighteen-week term. This suggests that the program of study was abbreviated and most probably part-time in nature.

This series of books supplies us with an early example of the Communist policy of incorporating political material in the teaching of mathematics. Illustrations of soldiers, rifles, and various machines of war are used to convey mathematical concepts. Word problems on the operation of subtraction consider the example of shooting down Japanese aeroplanes.[19] Another problem, concerning multiplication, states "If one communist soldier can kill three Japanese soldiers, five communist soldiers will kill how many Japanese?"[20] There is little doubt that the Japanese are the enemy! Practices of guerilla warfare, such as capturing enemy rifles, are emphasized in problem situations. Birth dates of Stalin, Marx, Lenin, Engles, and Mao are held up as references around which various chronologies are established,[21] and the Communist Party is in frequent mention. Even public sanitation

---

[19]Primary Arithmetic, vol. II, p. 24.

[20]Primary Arithmetic, vol. II, p. 35.

[21]Primary Arithmetic, vol. III, p. 25.

principles are focused upon in problem schemes involving the killing of flies.[22] The scope and sequence of the mathematics considered is as follows:

Book		Topics and Comments
I	1.	Understanding of the concepts; big, little; long, short; light, heavy; thick and thin.
	2.	Counting of sets of objects numbering from one to twenty. A set approach is employed and finger counting encouraged.
	3.	Finding the sum and difference of two single digit numbers using sets. Subtraction is presented as complementary addition.
	4.	Familiarization with units of money.
II	1.	Zero is introduced as an additive identity.
	2.	Addition of a two-digit addend and a one-digit addend with sum not exceeding nineteen.
	3.	Subtraction of a one-digit subtrahend from a two-digit minuend resulting in a one-digit difference.
	4.	Measurement of cloth.
	5.	Sum of three single-digit addends.
	6.	Multiplication introduced as repeated addition.
	7.	Addition of two double-digit addends with sum not exceeding thirty.

---

[22] Primary Arithmetic, vol. II, p. 35.

8. Subtraction involving a double digit subtrahend and a double digit minuend.

9. Multiplication facts established for numbers with products less than twenty.

III. 1. Review of previous mathematical knowledge.

2. Formation of number sentences, the symbol "=" introduced.

3. Division introduced as the inverse of multiplication.

4. Complete multiplication facts for two, three, four, and five established.

5. The formation of numerals representing numbers up to nine hundred considered and the concept of decomposition of a number introduced.

6. Consideration of units of weight.

7. Addition of two and three-digit addends.

8. Subtraction of two three-digit numbers.

9. Multiplication facts for six, seven, eight, nine established.

10. Algorithm for division developed.

11. Multiplication by zero considered.

IV. 1. Practice on telling time.

2. Division with a remainder introduced.

3. Multiplication involving a two-digit and a one-digit number.

4. Linear measure (decimal units).

       5.   Addition and subtraction of one place decimal fractions.

       6.   Work with measure of weight (rice).

       7.   Work with currency as a decimal system.

V.   1.   Writing of four place numerals.

       2.   Addition and subtraction of two three place digits.

       3.   Development of algorith for the multiplication of two double-digit numerals.

       4.   Oral exercises.

       5.   Practice with mixed operations.

       6.   Learn the number of days in the month, the months of the year.

       7.   More work on land and grain measure.

VI.   1.   The writing and understanding of five place numbers.

       2.   Practice with the four operations involving five place numerals.

       3.   Drill on solution of word problems.

Work with common fractions is absent from this program. Decimal fractions and their associated operations evolved from problems concerning various systems of Chinese measurement, all decimal in nature. The process of re-teaching the basic operations for each decimal grouping, i.e., tens, hundreds, thousands, appears pedagogically inefficient. Utilitarian needs, immediate and obvious, appear to be the guiding force behind this curriculum.

Yenan Accomplishments

By October 1944, there were 555 men-pan schools operating in the Communist areas in addition to the 526 existing regular schools,[23] which by now had also simplified their curricula. Thus the Communists in their pre-1949 educational experience had succeeded in devising educational techniques and curricula that were both popular and practical in their inception and appeal.[24] Much of the formalism of the usual Chinese school curriculum had been eliminated. In building a Chinese Communist community in Yenan, priority was given to the Chinese language and mathematics. While this decision was prompted by social needs, it also had political ramifications. Competence in language and mathematics was essential for the Communist Party's survival. Literacy was necessary for the effective reception of political indoctrination and mathematical knowledge requisite for industrial achievement. Military and political training, while not new dimensions in Chinese education, were now fundamental. Increased emphasis on production and the resulting closer association of students and workers led to a reduction of student disdain for manual work. In

[23]Seybolt, op. cit., p. 274.

[24]For a further discussion of Yenan education see Mark Selden, "Yenan Communism: Revolution in the Shensi-Kansu-Ninghsia Border Region 1927-1945" (unpublished doctoral dissertation, Yale University, 1967).

Yenan a system of education began to evolve that would
have satisfied the proponents of "practical education,"
who went unheard in the twenties. This system would continue developing after "liberation."

Mathematics Education in the Reconstruction of the School System in the People's Republic of China: 1949-1957

Withdrawal of the Nationalist government to Taiwan
in 1949 placed the Chinese Communist Party (C.C.P.) in
complete control of the mainland. Communist experience
at governing and providing for social needs, such as
education, although previously limited to small-scale
rural situations, provided models for the new government
to follow. Pragmatic, flexible standards reminiscent of
the Yenan Period were used in setting administrative
policy. Education was to be established upon a social-
economic basis, one relevant to the Chinese proletariat.
In this area, the tasks confronting the new government
were enormous: 80 per cent of the population was illit-
erate, no more than 40 per cent of the school population
could be accommodated by existing facilities,[25] textbooks were lacking, and in general the necessary enthusiasm for successful reforms would be difficult to muster.
Privations resulting from years of continued warfare,

----

[25]O. Fisher, "Education in Communist China,"
School and Society (June 20, 1959), pp. 302-305.

compounded by a rampant post-war inflation, had greatly demoralized the population. Wartime demands for educated personnel had thinned the ranks of teachers. Failure of the Kuomintang to assert a decisive authority and reunite China after the defeat of the Japanese alienated many former supporters of Chiang's government. Communist doctrines, anti-Kuomintang, and opposed to the capitalistic practices that had assisted in disrupting Chinese society, found a ready reception among this class of disillusioned intellectuals. This group now lent their expertise to the rebuilding of the new Chinese state.

Education, the primary molding force in Communist indoctrination, was given prompt attention by the Party's administrators. On September 21, 1949, the first plenary session of the Chinese People's Political Consultative Conference met to decide on specific governmental policies. The delegates adopted a Common Program or Constitution in which provisions for education were clearly stated:

> Article 41. The culture and education of the People's Republic of China are now democratic, that is, national, scientific, and popular. The main tasks for raising the cultural level of the people are: training of personnel for national construction work; liquidating of feudal, comprador, Fascist ideology; and developing of the ideology of serving the people.
>
> Article 42. Love for the fatherland and the people, love of labor, love of science, and the taking care of public property shall be promoted

as the public spirit of all nationals of the
People's Republic of China.

Article 43. Efforts shall be made to develop
the national sciences to place them at the
service of industrial, agricultural, and national defense construction. Scientific dis-
coveries and inventions shall be encouraged
and rewarded, and scientific knowledge shall
be popularized.

Article 44. The application of a scientific,
historical viewpoint to the study and interpretation of history, economics, politics,
culture and international affairs shall be
promoted. Outstanding works of social science
shall be encouraged and rewarded.

Article 46. The method of education of the
People's Government shall reform the old educational system, subject matter, and teaching
methods systematically according to plan.

Article 47. In order to meet the widespread
needs of revolutionary work and national construction work, universal education shall be
carried out. Middle and higher education shall
be strengthened; technical education shall be
stressed; the education of workers during their
spare time and the education of cadres who are
at their posts shall be strengthened; and revolutionary political education shall be accorded to
young intellectuals and old-style intellectuals
in a planned and systematic manner.[26]

These articles reflect the educational philosophies of

Mao Tse-tung and the pioneering experiments of the Yenan

Period. Defined in this declaration are three themes

around which educational activities in the People's

Republic would center--nationalism, popularism, and

scientism. These themes immediately became incorporated

[26]R. F. Price, Education in Communist China (New
York: Praeger Publishers, 1970), pp. 29-30.

127

into educational reconstruction. This phase of reconstruction passed through two periods: (1) Period of
educational rehabilitation and consolidation, 1949-1952;
(2) Period of sovietization, 1953-1957. To develop an
understanding of the educational advances made in Communist China during the years 1949-1957, each of these
periods must be considered separately.

Period of Educational Rehabilitation and Consolidation: 1949-1952

With the official establishment of the People's
Republic of China on October 1, 1949, a new Ministry of
Education was formed and placed under the jurisdiction
of the Committee on Cultural and Educational Affairs.
The Ministry's initial concern was to keep the 310,000
schools it inherited functioning; therefore, immediate
reforms intended for this purpose were not exceptional
in themselves. Mindful of Mao's dictates to systematically reform the old educational system, China's new
educational administration attempted to incorporate the
experiences of the Liberated Areas with Soviet educational theories. Russia now became China's educational
exemplar.

Reforms Designed to Increase Enrollments

Immediate attention was given to increasing school
enrollments. Class scheduling was altered to accommodate

the greatest number of students possible by allowing
three shifts a day. As in pre-liberation days, classes
met six days a week. Facilities for part-time and self
study were expanded. School curriculum deviated little
from previous Nationalist requirements. Primary school
studies included civics (adapted to Communist theories),
arithmetic, nature study, arts and crafts, social
studies, and physical activities (games) for the first
four years. Plans for the last two years of primary
school studies replaced arts and crafts with manual
work, nature study with general science, and social
studies with hygiene and sanitation principles. Total
curricula time was shorter than that of previous Nationalist programs. Middle school reforms were also minor.
Subjects studied at the lower level included languages--
Chinese and English or Russian; mathematics--arithmetic,
elementary algebra, and elementary geometry; social
studies--world history, geography, and Marxist theories
for at least two hours a week; natural science--general
science and biology; manual arts; and physiology, hygiene,
and physical activities. Senior middle school still
operated a multi-track program, providing streams for
further academic pursuits, modern agriculture, business
and commercial training, and teacher education. Studies
in Chinese language, mathematics, social science, and
physical education constituted a required core. Upon

completion of core work, students were free to special-

ize.[27] Under the reforms, science majors were exempted

from advanced courses in history and geography and had

their language requirements shortened. Similarly, arts

majors no longer had to study chemistry and physics,

and took fewer courses in mathematics than had previously

been required.[28] Scouting activities were condemned as

reactionary and removed from the middle school curriculum.

The Children's Corps of China (C.C.C.) was founded to

replace the scouts. It was felt that the C.C.C. offered

a child more freedom to develop as an individual and

provided encouragement for academic work, particularly

mathematics and science.[29] The Ministry issued a series

of educational reforms in the summer of 1950: "Decision

on Textbooks for Middle and Primary Schools for the Autumn

of 1950," "Temporary Plan of Pedagogy for Middle Schools,"

and a "Temporary School Calendar for Middle Level

Schools." Innovations in mathematics teaching included

the dropping of analytic geometry from the senior middle

school program, simplification of the solid geometry

---

[27]C. S. Bo, Programs of New Education (Shanghai: People's Publishing House, 1949), p. 211.

[28]Derk Bodde, Peking Dairy 1948-49: A Year of Revolution (New York: Fawcett Premier Books, 1967), p. 249.

[29]Edmund J. King, Communist Education (New York: Bobbs Merrill Co., 1963), p. 275.

course, and an increase in scheduled classroom time
devoted to mathematics review. Quantitative results of
the government's educational reforms soon became obvious:
between the years 1949 and 1952, the number of students
in China's primary and middle schools doubled![30]

Emphasis on Accelerated and Informal
Education

With the continuity and expansion of the educational process assured, reformers now turned their attention to the structure of the school system. Duration of
the time period spent in schooling became an issue for
criticism. Twelve years was thought to be too long a
period of non-productive activity. A six-four system
was suggested to replace the twelve-year system. Rather
than reduce the length of secondary schooling, Chinese
educators under Soviet influence elected to change to a
five-year unified primary program. Justification for
this innovation was given by Liu Shih, Director of the
Supervisory Department of the Ministry of Education in
1951:

> In the old educational system the primary
> schools were divided into two classes. The
> first was of four years and the second of two
> years. The two together provided a full
> period of primary schooling. There are more

[30] R. F. Simpson, "The Development of Education in
Mainland China," Phi Delta Kappan (December, 1957), pp.
84-93.

than 440,000 primary schools in China of which
only 10 per cent provide a complete primary
schooling. They are all concentrated in cities
and big towns. That is to say, the children of
the vast peasant class had none or very little
opportunity to receive a complete primary education.[31]

Thus both production and proletarian demands were
reflected in this decision.

The First National Conference of Middle Schools
meeting in March of 1951 asserted the retention of the
middle school program to six years. The conference urged
that programs of extracurricular activities be utilized
by schools to supplement and further academic studies.[32]
Necessity for academic streaming at the senior middle
school level was replaced by the establishment of large
numbers of specialized middle schools, most of which were
technical in nature. This polytechnic approach to education paralleled Soviet experience of the early twenties.
Factories, middle schools, and army units were advised to
establish their own short-term middle schools for workers
and peasants. Short-term schools offered abbreviated
three-year middle school equivalency programs. Such an

---

[31]Lui Shih, "China's New Educational System,"
People's China (December 1, 1951), p. 7.

[32]Richard Arens, "The Impact of Communism on
Education in China, 1949-50" (unpublished Ph.D. thesis,
University of Chicago, 1952), p. 138.

action on the government's part rested more in psychological appeasement than academic soundness. Now, theoretically, a middle school education was available to
almost everyone.

Rural spare-time educational facilities were
expanded. Standards in these schools were quite flexible and the curriculum basic. The government's
"Directives Concerning the Development of Spare-time
Education for Peasants" of December 14, 1950, directed
that school work be tailored to specific needs of the
locale, but suggested that some studies focus on developing a written language proficiency, supplemented by
knowledge of current affairs, agricultural production,
and health and sanitation principles.[33] A later pronouncement of May 1951, "Temporary Regulations for the
Carrying-out of Spare-time Education for Workers and
Staff Members," was more explicit in its requirements
for two-year primary and middle school spare-time students. A general timetable for spare-time schools is
given in Table 9 below.

[33]Chi Tung-wei, Education for the Proletariat in
Communist China (Hong Kong: Union Research Institute,
1956), p. 10.

TABLE 9

GENERAL TIMETABLE FOR TWO-YEAR SPARE-TIME
SCHOOLS, CIRCA 1951[34]

| Year | Primary Class | | | | Secondary Class | | | |
| | I | | II | | I | | II | |
	Hr/ wk.	Total	Hr/ wk.	Total	Hr/ wk.	Total	Hr/ wk.	Total
Chinese	4	160	4	160	2	80	2	80
Arithmetic	2	80	2	80	2	80	2	80
Common Knowledge	_	___	_	___	2	80	2	80
	6	240	6	240	6	240	6	240

Trained teachers, if available, were drafted into service.
Most often, however, a "knowledgeable" person would be
selected from the ranks to serve as an instructor. When
regular textbooks were in supply they were used. In
other cases, teachers wrote their own materials or relied
on the use of official party documents or newspaper
articles as reading matter. Yenan-like improvisation
was necessary. Graduation from the program required a
reading knowledge of 2,000 Chinese characters for workers
and 1,500 characters for peasants and a working knowledge
of arithmetic.[35]

Table 10 lists the courses of study for a regular
middle school of the period.

---

[34]Ibid., p. 13.

[35]Elsie Fairfax-Cholmeley, "Education in China,"
New World Review (August, 1956), p. 37.

TABLE 10

CURRICULUM OF GENERAL MIDDLE SCHOOLS FOR SOUTH
KIANGSU PROVINCE, 1951 (HOURS/WEEK)[36]

	Junior Middle School			Senior Middle School		
Year	1	2	3	4	5	6
Subject						
Politics	2	2	3	3	3	3
Chinese Literature	6	6	6	5	5	5
Arithmetic	4	–	–	–	–	–
Algebra	–	4	–	–	–	–
Plane Geometry	–	–	4	–	–	–
Trigonometry	–	–	–	2	–	–
Solid Geometry	–	–	–	–	4	–
Advanced Algebra	–	–	–	–	–	4
Russian or English	4	4	4	4	4	4
Geography	2	2	2	2	2	2
History	3	3	3	3	3	3
Chemistry	–	3	–	–	4	–
Physics	–	–	3	–	–	4
Physiology	–	1	–	–	–	–
Biology	–	–	–	3	–	–
Zoology	2	–	–	1	–	–
Botany	2	–	–	1	–	–
Fine Arts	1	1	1	1	–	–
Music	1	1	1	1	1	1
Physical Ed.	2	2	2	2	2	2
Total hours	29	29	29	28	28	28
Credit value	Total of 32			Total of 35		

In order to meet the new demands put upon the
system, teacher training programs were also revised.
Spare-time and concentrated programs of study augmented
the traditional normal study programs. Curricula, with

---

[36]Anthony Chen, "The Philosophy of Communist China
as Applied to Secondary and Higher Education" (unpublished
M.A. Thesis, De Paul University, Chicago, 1956), p. 91.

the exception of required political studies, remained basically the same as before liberation. A radical change did take place, however, at the university level, which affected schools of education. Following its new polytechnic philosophy, the government took over all private universities, consolidated some, and partitioned others into several different schools. Previous departments or schools of education of a university now became independent teachers colleges or normal universities. One example of this phenomenon is given by the East China Normal University's absorption of the College of Letters and Science of Kuanghua University and Great China University and the education departments of five other universities.[37] By the application of such a policy, foreign influence was eliminated and extreme specialization could be achieved. This principle became known as Chuan-yeh, and was explained by Vice Minister of Education Tseng Chao-lun in a 1952 speech:

> Chuan-yeh does not denote a type of school nor a type of department in a school. What it does denote is simply this: that everything taught in the specialized school of higher learning should be directed towards a clearly defined specialization in one or more subjects. This is the purpose behind the radical reshuffling of colleges and departments of schools of higher education.[38]

[37]Chen Hsuan-shan, "Training Teachers for Middle School," China Reconstructs (September, 1956), p. 19.

[38]Translated in China News Analysis, No. 81, April 29, 1955, pp. 5-6.

The Effects of Educational Rehabilitation
on Mathematics Teaching

Mathematics teaching reforms during the period
1949-1952 were slight. Revision of specific disciplines
had not yet been given priority by Chinese educators.
In middle school mathematics, the level of rigor was
slightly lessened and the scope of possible study
reduced by the exclusion of analytic geometry. In 1952,
the Ministry of Education published a new "Syllabus for
the Teaching of Mathematics in Middle Schools," which
resulted in the publication of a corresponding series
of texts. The contents of the new books promoted a
socialist outlook; problem situations dealt with production norms and national accomplishments. Mathematical
theory was associated with practical applications

Period of Sovietization: 1953-1957

Enlarged school enrollments, while desirable, also
produced some disadvantages. Unfulfilled educational
demands of industry soon made the deficiencies of the
school system apparent. Sufficient numbers of experienced teachers were lacking, and those available were
weak in subject background. As a result of the poor
educational showing, attempts at the five-year primary
program were abandoned late in 1953. In the beginning of
the same year, the government had launched its First Five
Year Plan. Under this plan, heavy industry was to be

developed, providing a base for the industrialization
which was to convert China "from a backward agricultural
country into an advanced, socialist, industrial state."
All agencies of the Chinese society, including education,
were to contribute to this effort. During the duration
of the plan, schools were expected to produce a quarter
of a million engineers and technicians.[39] To realize
this quota, reforms had to be undertaken and teaching
standards revised. On May 17, 1953, Mao officially
called for these reforms:

> The so called teaching reform is the reform of
> educational contents and teaching methods.
> Consequently, it is necessary to revise and
> compile teaching materials and to compile
> pedagogical methods.[40]

Soviet assistance in the required revision and compilation of materials was heavily relied upon. Guided by the
contemporary slogan "learn from the advanced experience
of the Soviet Union," the Chinese emulation of Soviet
educational practices now far exceeded the extent to
which American influence had predominated in the thirties.
This emulation was strengthened by (1) mass translation
of Russian texts and teaching outlines, (2) large numbers
of Chinese students being sent to Russia to study, and

---

[39]Simpson, op. cit., p. 88.

[40]"Chronology of the Two-Road Struggle on the
Educational Front in the Past Seventeen Years," translated
in Chinese Education (Spring, 1968), p. 14.

(3) the influence of Soviet educational advisors.

Adoption of Soviet Teaching Plans and Texts

In 1953, the Ministry of Education promulgated teaching plans for middle schools in which comparative Soviet middle school teaching outlines were adopted in toto. A series of texts were compiled that followed these outlines exactly, and in turn teachers coordinated their classroom lesson plans to conform to the new requirements. Teaching outlines became dogmatic guides for improved instruction. The Soviet policy of organizing study groups or teams for research and teaching, used so effectively in the Yenan period, was followed in all schools. Research and teaching group membership united teachers involved in the teaching of a specific discipline; at the primary level, membership was universal. Collective efforts in preparing lesson plans insured adherence to the latest governmental political and pedagogical directives. At higher levels of education, including normal schools, "research" efforts were directed at translating Russian texts and teaching materials and adapting them for Chinese audiences. The Russian language was made a required subject in all schools above the primary level, and in 1954 the teaching of English was suspended. Between 1952 and 1956 an

estimated 1,400 textbook titles were translated.[41] Most
of these works were technical in nature and intended for
the vocationally-oriented schools. Many applied, however, to the teaching of general primary and middle
school subjects, especially mathematics.

Soviet Mathematics Texts

Soviet educational literature contains a rich
variety of works designed to improve mathematics teaching. Much of this material now found its way into the
bookshops and normal schools of the People's Republic.
Books such as A. S. Pchelko's Methods of Teaching Arithmetic in Elementary School, Y. Liapin's Middle School
Mathematics Pedagogy, and Chichigin's Middle School
Arithmetic Teaching Methods became standard references
for Chinese teachers. Such varied aspects of mathematics education as the psychology of geometry learning,
the giving of examinations, and the grading of homework
were now documented from Soviet sources. Many of these
works were not published under the Soviet authors'
names, but were compiled and summarized by Chinese
writers or writing groups. Examples of these compila-
tions are: Reference Material on Mathematics Teaching,
by the Nanking Teachers College writing group, and

---

[41]Price, op. cit., p. 102.

Explanation of Junior Middle School Algebra, by Cheun
Tung-ping. Although specific middle school texts, such
as Chichigin's Geometry, were completely translated and
published in Chinese, they were intended for teacher and
textbook writers' reference, rather than as instruments
of classroom instruction. Middle school textbook revision was undertaken in 1955 to conform to the Ministry's
1954 syllabus for mathematics teaching. Chinese authors
relied heavily on Soviet material, sometimes incorporating
the contents and methods of several Russian texts into
one of their own. The resulting series of mathematical
texts was considered an improvement over the 1952 edition, but it still came under criticism. Many educators
felt that the books contained an excessive repetition of
topics and that content presentation did not associate
mathematical knowledge with the technical needs of the
state. A revision of the syllabus for the school year
1955-56 appeased the protagonists of labor orientated
mathematics. The new outline added practical surveying
techniques to the middle school curriculum, stressed
methods of approximation and curve sketching, and
increased the proportion of production related exercises.[42]

---

[42]Lee Yee-chung, "Expressing My Opinion on Party
Policy [in Mathematics Education]," Chung-hsueh Shu-
hsueh (January, 1959), pp. 2-4.

A detailed middle school mathematics syllabus is available for the school year 1956-57.[43] An outline of this syllabus is as follows:

Seventh Grade

    <u>Arithmetic</u> 6 hours/week   Total 204 hours

1.	Work with whole numbers	33 hours
2.	Preparation for fractions	20 hours
3.	Fractions	67 hours
4.	Decimals	33 hours
5.	Percentage	12 hours
6.	Ratio	27 hours
7.	Survey practice	6 hours
8.	Review	6 hours

Eighth Grade

    <u>Algebra</u> 4 hours/week   Total 136 hours

1.	Equations and Functions	20 hours
2.	Rational numbers	24 hours
3.	Equations with integral coefficients	46 hours
4.	Factoring of Polynominal equations	22 hours
5.	Rational equations	20 hours
6.	Review	4 hours

    <u>Geometry</u> 2 hours/week   Total 68 hours

1.	General introduction	16 hours
2.	Triangles	42 hours
3.	Survey practice	6 hours
4.	Review	4 hours

Ninth Grade

    <u>Algebra</u> 3 hours/week   Total 102 hours

1.	Rational expressions	22 hours
2.	Ratio	6 hours
3.	First degree equations (one variable)	

---

[43]Ministry of Education, People's Republic of China, "Middle School Mathematics Teaching Outline: 1956-57," <u>Shuxue Tongbao</u> (August, 1956), pp. 24-39; for complete outline see Appendix C.

Ninth Grade (Continued)

    4. First degree equations
        (two variables)   22 hours
    5. Square roots   12 hours

Geometry 3 hours/week   Total 102 hours

    1. Parallel lines   16 hours
    2. Quadrilaterals   24 hours
    3. Circle   30 hours
    4. Inscribe and circumscribe
        circles in and about tri-
        angles and quadrilaterals 10 hours
    5. Survey practice   8 hours
    6. Review   14 hours

Tenth Grade

Algebra 4 hours/week (first semester)
       3 hours/week (second semester)
                     Total 120 hours

    1. Powers and Roots   36 hours
    2. Factoring of second degree
        equations   46 hours
    3. Functions and Graphs   17 hours
    4. Solutions for systems of
        equations in two variables 18 hours
    5. Review   3 hours

Geometry 2 hours/week (first semester)
       3 hours/week (second semester)
                     Total 84 hours

    1. Similarity   33 hours
    2. Trigonometric function of
        acute angles   10 hours
    3. Properties of triangles and
        circles   10 hours
    4. Area of polygons   12 hours
    5. Survey practice   6 hours
    6. Review   3 hours

Eleventh Grade

Algebra 2 hours/week   Total 68 hours

    1. Series   18 hours
    2. Advanced theory of powers 8 hours
    3. Exponential and logarithmic
        functions   40 hours
    4. Review   2 hours

Eleventh Grade (Continued)

Geometry 2 hours/week   Total 68 hours
Plane Geometry

   1. Regular polygons   12 hours
   2. Circumference and area of
       circle   11 hours

Solid Geometry

   1. Straight line and plane   43 hours
   2. Review   2 hours

Trigonometry 2 hours/week   Total 68 hours

   1. Values of trigonometric
       functions for angles
       between 0 and 360
       degrees   18 hours
   2. Radian measure   3 hours
   3. Trigonometric functions
       of any angle   10 hours
   4. Trigonometric formulae   20 hours
   5. Logarithms of trigonometric
       functions   36 hours
   6. Rectangular pyramids   8 hours
   7. Survey practice   4 hours
   8. Review   8 hours

Twelfth Grade

Algebra 2 hours/week   Total 68 hours

   1. Combinations, permutations
       and the Binomial Theorem   12 hours
   2. Complex numbers   10 hours
   3. Inequalities   22 hours
   4. Higher degree equations   12 hours
   5. Review   12 hours

Geometry 2 hours/week   Total 68 hours

   1. Straight line and plane   6 hours
   2. Polyhedra   30 hours
   3. Solids of revolution   20 hours
   4. Review   12 hours

Trigonometry 2 hours/week   Total 68 hours

   1. General trigonometry   22 hours
   2. Inverse trigonometry func-
       tions   10 hours
   3. Trigonometric equations   16 hours
   4. Survey practice   4 hours
   5. Review   16 hours

This middle school mathematics curriculum provides some
indication of the rigor, scope, and style of mathematic
studies during this period. The apparent redundancy of
topics and the classical nature of the material presented
is obvious. At a later time, this syllabus would be
referred to as an example of academic excellence. For an
examination of the complete primary and middle school curricula of this period, see Tables 11 and 12 below.

During this same period, several journals intended
for mathematics teachers and fashioned after the Soviet's
Matematika v Shkole and Mathematiches Koye Prosveschenia
were begun. Two of the more influential of these publications were Chung-hsüeh Shu-hsüeh and Shuxue Tongbao.[44]
Their pages provided a forum for Chinese teachers to share
experiences; since many of their articles were translations from Russian sources, they supplied an additional
avenue of information on Soviet teaching practices.

Increased Emphasis on Teacher Training

Teaching standards improved as a result of the
influx of Soviet teaching references and the efforts of
teaching and research groups. Classrooms were badly
overcrowded, with perhaps forty to fifty students to a
teacher at both the primary and secondary levels.
Despite an eight-hour, six-day-a-week work load, teachers

---

[44]Middle School Mathematics, Huanan Pedagogic
Institute (monthly); Mathematics Bulletin, Chinese Mathematical Society, Peking (monthly).

were expected to be involved in students' extracurricular activities, political studies, and the work of research groups. Between the years 1951-1956 approximately 300,000 primary teachers were graduated from normal schools, and an equally large increase took place in the middle school labor force.[45] Although the numbers of modern, trained teachers were increasing, the general calibre of the teaching profession remained low. In 1956, 39 per cent of all primary school teachers had less than a lower middle school background, 90,000 middle school teachers lacked a college degree, and 70 per cent of the faculty involved in higher education had never done post-graduate work.[46] Teacher-training programs were revised in 1956 and in-service education programs made available to working teachers. Correspondence courses, spare-time courses, and refresher courses provided during leaves of absence were also made available to teachers with weak academic background. Between 1950 and 1958, 583 Soviet educational advisors worked in the universities and colleges of the People's Republic of China.[47] While no specific statistics are available, it

---

[45]Simpson, op. cit., p. 90.

[46]Theodore Chen, Teacher Training in Communist China, Studies in Comparative Education Series, No. OE 14058 (Washington, D. C.: U. S. Office of Education, 1960), p. 29.

[47]Price, op. cit., p. 102.

is almost certain that some of this number were involved
in normal school teaching and curricular reforms.

Selected teacher-trainees were sent to the Soviet
Union for higher studies. The eventual number of Chinese
students to be trained in Russia under this exchange
program would total 61,000. This number far exceeds the
35,931 Chinese students educated in America between 1905
and 1952.[48] It provides an additional indication of the
extent of Soviet influence as compared to the previous
educational influence exerted on China by the United
States in the first half of the century.

TABLE 11

PRIMARY SCHOOL TIMETABLE
1956 (HOUR/WEEK)[49]

Subject	1	2	3	4	5	6	Total Hours
Chinese Language	12	12	12	12	9	9	2,244
Arithmetic	6	6	6	7	6	5	1,224
History	–	–	–	–	2	2	136
Geography	–	–	–	–	2	2	136
Natural Science	–	–	–	–	2	3	170
Physical Education	2	2	2	2	2	2	408
Music	2	2	2	1	1	1	306
Art	1	1	1	1	1	1	204
Handicrafts	1	1	1	1	1	1	204
	24	24	24	24	26	26	5,032

_____

[48]Y. C. Wang, Chinese Intellectuals and the West,
1872-1947 (Chapel Hill: University of North Carolina
Press, 1966), p. 511.

[49]UNESCO, World Survey of Education, Vol. II
(Paris, 1958), p. 255.

TABLE 12

CURRICULUM FOR GENERAL MIDDLE SCHOOL
1956-57[50]

Subject	Junior Middle School			Senior Middle School		
	1	2	3	4	5	6
Chinese Language	3	3	2	1	1	1
Chinese Literature	6	6	5	4	4	4
Arithmetic	6	–	–	–	–	–
Algebra	–	4	3	4/3	2	2
Geometry	–	2	3	2/3	2	2
Trigonometry	–	–	–	–	2	2
Chinese History	–	3	3	–	3	3
World History	–	–	3	–	–	–
Contemporary History	–	–	–	3	–	–
Political Education	–	–	2	–	–	–
Social Science	–	–	–	–	1	–
Constitution	–	–	–	–	–	1
Geography	3	2/3	3/2	2	2	–
Botany	2	3/0	–	–	–	–
Zoology	–	0/3	2	–	–	–
Human Anatomy	–	–	–	2	–	–
Darwinism	–	–	–	–	2	–
Physiology	1	–	–	–	–	–
Physics	–	3/2	2	3	3	5/4
Chemistry	–	–	2/3	2	2	3
Foreign Language	–	–	–	4	4	4
Physical Education	2	2	2	2	2	2
Advanced Drawing	–	–	–	1	1	1
Laboratory & Practical Training	2	2	2	2	2	2
Music	1	1	1	–	–	–
Total	30	32	32	32	33	32

A Dependence on Examinations

It is interesting to note that although the Communists condemned the elitist educational practices of

[50]Evelyn Harner, Middle School Education as a Tool of Power in Communist China (Santa Barbara, Calif.: General Electric Co., 1962), pp. 26-27.

traditional China, including the civil service examinations, they themselves now established an educational system that stressed examinations. Ascendency at every level of the educational ladder depended on passing examinations. Teaching outlines specified the numbers of quizzes and examinations to be taken by individual classes. Passing from primary to middle school was determined by two examinations, one to qualify for promotion from the primary level, the other to ascertain acceptance into middle school. Similar situations existed at the junior-senior and middle school-university transitions. As a result, students and teachers became examination-oriented. Many schools adopted a Soviet five-point grading system in place of percentage ratings. The following high school entrance examination typifies the style of mathematics testing popular at this time in the People's Republic.

MATHEMATICS ENTRANCE EXAMINATION FOR
SENIOR MIDDLE SCHOOL (SUMMER 1955)[51]

Part I (7 points each)

1. Find an equation whose roots are the square of the roots of
    $$x^2 - 3x - 1 = 0$$

---

[51]"Summer 1955 Senior Middle School Entrance Examination in Mathematics," Shuxue Tongbao (August, 1955), pp. 47-48.

2. Find an expression for the interior angles of an isosceles triangle whose side is four times the length of the base.

3. Given a square right pyramid whose base has a side of length a. If the angle the pyramid's face makes with the base is 45°, what is the altitude of the given solid?

4. Two planes intersect in space. From a point in each plane a normal is constructed. The two normals to the given planes are coplanar and meet. What is the angle of their intersection called?

Part II (18 points each)

1. Find the coefficients b, c, and d such that $x^2 + bx + cx + d$ can:
   a. be divided evenly by x-1
   b. be divided by x-3 with remainder 2
   c. be divided by x + 2 and x - 2 and have the same remainder in both cases.

2. Given the triangle ABC, circumscribe a circle about the triangle, from a point D on side AC draw a line perpendicular to side AB and extend it to intersect the extension of BC at F. The constructed line intersects the circle at G.
   Prove: $(EG)^2 = EF \cdot ED$

3. Solve for x:
   $$\cos 2x = \cos x + \sin x.$$

4. Given a triangle with a perimeter of 12 ft. and area 6 square feet.
   Prove: The given triangle is a right triangle and that the length of its sides are 3, 4, and 5 feet.

Perhaps most reminiscent of the traditional civil service examinations were the mathematical competitions begun in several major Chinese cities in 1956.[52] These

[52] See John De Francis, "Mathematical Competitions in China," The American Mathematical Monthly (1962), 52: 251-255.

examinations, patterned after the Soviet Mathematical Olympaids,[53] were intended to stimulate student interest in mathematics and science and to locate budding mathematical talent that could be developed by the state. The design and rationale of this scheme clearly follows that employed by the Hanlin Academy in its grooming of literary scholars. Now, however, the ultimate objective is technical proficiency rather than scholarly humanism.

Objections to Sovietization

Complete emulation of Soviet practices and the adopting of Soviet theories alarmed some educators. Students and teachers were greatly burdened by the demands of the new curricula. Extreme specialization had weakened the overall importance and applications of many disciplines. One source cites an institution of higher learning where mathematics teaching had become fragmented into nineteen branches.[54] Hastily translated Russian texts often obscured knowledge. In May of 1956, the government of the People's Republic, satisfied with its progress towards reconstruction, allowed public criticism

---

[53]For example, from these examinations see D. O. Shklarsky, N. N. Chentzov, I. M. Yaglom, The U.S.S.R. Problem Book: Selected Problems and Theorems of Elementary Mathematics (San Francisco: W. H. Freeman and Co., 1961).

[54]Leo Orleans, Professional Manpower and Education in Communist China (Washington, D. C.: National Science Foundation, 1961), p. 93.

of its policies by declaring the "Hundred Flower Campaign." Under the slogan "let one hundred flowers bloom and one hundred schools of thought content," doubts concerning the accomplishments of the new state were voiced. Particularly in the educational realm, strong criticism was directed at Sovietization:

> During the past few years, the fundamental feature [in education] has been the mechanical copying of Soviet experiences. There has been a strong tinge of doctrinairism. Up to the present, Chinese instructors of higher education are still using only Russian textbooks on education. No textbooks on education have been written and published by ourselves to suit the actual conditions of China . . .[55]

Singled out for special objection was the "three copy method" used in instruction: the teacher copied his notes from a Russian source; he then copied them onto the board and the students copied them into their notebooks. This cycle was purely mechanical and limited as a learning process. If a student were to question any of the content, he was challenged with "This teaching material originates from the Soviet Union."[56] The quality of education was questioned, particularly that supplied by part-time or people's schools. One of the government's

---

[55]Chu Yu-hsien, Professor of Education, East China Normal University, quoted in Roderick MacFarquhar, The Hundred Flowers Campaign and the Chinese Intellectuals (New York: Fredrick Praeger, Inc., 1960), p. 91.

[56]Dick Wilson, Anatomy of China (New York: Weybright and Talley, Inc., 1966), p. 79.

model schools, the People's University, was singled out

for scathing comment:

> The People's University is a university in name
> only, and it resembles a secondary school in
> the content of its instruction and a primary
> school in teaching methods.[57]

> The People's University does not look like a
> school, but a large beehive of dogmatism. All
> it has done is to disseminate dogmatism.[58]

Objections to the government's educational poli-

cies did not result in sweeping reforms as the critics

had hoped. Instead, alarmed by the extent and vociferous-

ness of the intellectuals' attacks, the government set

out to silence the voices of dissent. Its reaction was

to undertake a "rectification" campaign in April of 1957,

through which misguided intellectuals were directed back

to the paths of party policy.

Educational Advances: 1953-1957

While the educational advances of this period may

have been questionable, one fact remains unchallenged:

Soviet influence had made a lasting impact on the

Chinese educational scene. In mathematics education,

two far-reaching innovations resulted: an instilling in

Chinese educators of the importance of studying the

psychological foundations of mathematics learning and a

---

[57]Orleans, op. cit., p. 93.

[58]Ibid.

153

revision of school mathematics texts along less classical
lines. For years, Soviet educational researchers had
been involved in studying the learning processes of children in regard to specific disciplines and in incorporating their findings into instructional strategies. Under
Soviet tutelage, the Chinese began their own work in this
field and established several experimental schools in the
People's Republic.[59] Soviet-influenced textbook revision
in 1953 and again in 1955 produced more readable and
better-organized texts from which students benefited. The
collective efforts of mathematics research groups assisted
in improving teaching standards, and the competitive
nature of the Mathematical Olympiad sparked a renewed
interest in the discipline for China's youth. However,
in reviewing this educational progress, party leaders
became concerned with the lessening of the role politics
was playing in education. On March 6, 1957, Mao issued a
statement within which he urged that political education
be stressed and the required number of class hours be
reduced. He believed that the teaching materials should
be revised to better reflect local conditions.[60] With
this pronouncement, the centralized control of education
in China was diminished.

[59]For a detailed discussion, see section on
psychological research below.

[60]"Chronology of the Two-Road Struggle," op. cit.,
p. 30.

Conclusions

Initial Socialist educational reforms in education,
while based on academic goals divergent from those established by the Nationalists, functioned in a similar manner,
both school structures and curricular emphasis differed
little. With the Yenan exile the Communists reformulated
their educational philosophy and pedagogy making a distinct
break with traditional thinking. Classroom content became
based on the solicited needs of the people. Mathematics
teaching centered on: account keeping, systems of weights
and measures, land mensuration and abacus drill. This
proletarian relevance was short-lived as "liberation"
forced the Communist Party to face the responsibility of
running a national school system. Now confronted with the
pressures of guiding the destiny of the Chinese nation,
the Communists reverted back to pre-liberation education
practices and resurrected some spectors of traditionalism--
scholarly elitism and a highly selective academic examination process.

CHAPTER IV

THE REORIENTATION OF MATHEMATICS EDUCATION
TO SATISFY THE NATIONAL GOALS OF THE
NEW CHINESE STATE

While the Communist government had succeeded in preserving and extending the inherited school system, it was guilty of perpetuating many of the deficiencies found in pre-liberation education. Schooling was still an elitist activity concerned to a large extent with university preparation. Teaching suffered from formalism, and a student's academic success was contingent upon examination performance. The structure and curricula of the new educational system were imported in toto from the Soviet Union. Soviet-inspired emulation resulted in the promotion of a Marxist-Leninist world outlook in the teaching of many subjects, but also perpetuated a cultural alienation of the Chinese student from the realities of his society. The value of this practice now became doubtful, and attempts were made to evolve an educational model more suited to China's economic and social needs.

The Period of Experimentation (The Great
Leap Forward in Education): 1958-1960

The year 1958 saw a great amount of confusion on the Chinese educational scene. A series of government directives were issued with the intent of furthering

politics, proletarianism, and production in education.
Publicly, primary emphasis appeared to be on production;
but political considerations, while not as obvious, were
paramount within the new reforms. The Communist Party
was to assert itself as the sole architect of educational
policy. Education was swept along in the grand designs
of the "Great Leap Forward," a period in which all
aspects of Chinese society, industrial, economic, and
cultural, were to advance greatly and narrow the developmental gap with the Western world.

"Redness and Expertness"

On January 31, 1958, Mao set in motion a revolution in education with his "Sixty Articles of Working
Methods" directive. In this document, he stressed the
importance of a "politico-ideological" basis for education: "There is no question that politics is unified
with economics and technique . . . This is called red
and expert."[1] Redness was to be acquired as a result
of increased political education in the schools. Expertness, while being acquired through regular academic
studies, would be accelerated by assuring a closer relationship between studies and physical labor directed

---

[1]Mao Tsê-tung, "Method of Work Draft," Current
Background, No. 888, "Chairman Mao on Revolution in
Education" (Hong Kong: United States Consulate General,
August 22, 1969), p. 8.

toward production. Mao was affirming his epistomological
philosophy at a most opportune time--to coincide with
the Great Leap Forward and the beginning of the state's
Second Five Year Plan. All students from the primary
through the university level were required to engage in
productive labor, establish factories, plant crops, and
in general become part of the national productive force.
Student-based labor, while contributing to production
quotas, also helped schools attain financial self-sufficiency and reduced a lingering burden on the state.
Contact between students and teachers and workers and
peasants, while performing physical labor, proletarian-
ized the intellectuals and intellectualized the proletariat. In March, a new pedagogical plan was issued by
the Ministry stipulating specific hours and quotas for
productive labor. Labor was no longer a theoretical or
token school subject as in the past; schools now became
affiliated with factories, mills, and communes. Official
party policy on this matter was given by Lu Ting-yi,
Director of Propaganda, in a speech entitled "Education
Must be Combined with Productive Labor." Lu advocated
the principles of "diligent in work and frugal in study"
and urged the establishment of secondary spare-time
agricultural schools. The general slogan of "more, better,
faster, and cheaper" applied to the industrial sector of
the economy was also adopted in education. Former

bourgeois policy whereby "education was run by experts" and "professors must run the schools" had to be changed.[2]

By the summer, many schools had established factories of their own. As in all government campaigns, popularized slogans were in abundance, one of which, "walking on two legs," perhaps best summarizes the prevailing attitude in the People's Republic at this time. "Walking on two legs" meant solving a problem by all available means. Thus the harnessing of student labor combined with raised industrial quotas would achieve increased production by two diverse means. Similarly, in education a leap was being realized by requiring accelerated building programs and timetables, and also by encouraging factories, mills, military garrisons, and farms to establish their own schools. Spare-time and half-study, half-work facilities were multiplied many times over due to the government's pronouncement that "every knowledgable person could teach." These institutions varied from the usual min-pan school to "red and expert universities." Renewed emphasis on spare-time education was intended to intellectualize the proletariat, providing a psychological appeasement for the masses' desire for education. Spare-time universities were far

---

[2]Lu Ting-yi, "Education Must Be Combined with Productive Labor" in Stewart Fraser, Chinese Communist Education (Nashville: Vanderbilt University Press, 1963), pp. 283-300.

from being universities. Evelyn Harner, in her study on
Communist education, provides some statistics on the background of students in the Communist Labor University--
almost 70 per cent of the students had only a primary
school education.[3] During this period of rapid reform,
the Ministry of Higher Education was abolished, and its
functions transferred to the Ministry of Education.

Mathematics in Spare-time School Programs

Although in this informal school system, curricula
varied from institution to institution, major courses
were similar to those included in previous part-time
programs: Chinese language, mathematics, politics, and
production experience. Standards were expected to be
equivalent to those of the regular school system. Peking
Normal University published a recommended teaching outline for arithmetic for a one-year spare-time course for
farmers and soldiers. The list of topics to be studied
included:

    I.   Integers
         1. Character and rules of calculation.
         2. Calculations involving mixed operations.
         3. Abacus practice.
   II.   Fractions, decimals, percentage and propor-
         tion.
  III.   Simple statistical and accounting principles.

----

[3]Harner, op. cit., p. 22.

IV. Survey of area, volume and distance.
V. Construction of graphs and charts.[4]

Some schools such as the Tsinchow Textile and
Machine Manufacturing School had ambitious spare-time
mathematics study programs (see Table 13). Their level of
rigor exceeded that required of regular middle schools.

Within a few months from the start of the expanded
part-time educational programs, tremendous advances on
the education front were announced. By June, it was
claimed that 1,240 districts had achieved universal
primary education and that 90 million people were attend-
ing literacy classes.[5] Two hundred twenty-four spare-
time colleges had been established to absorb the flow of
candidates desiring higher education. By November, the
government could document the Leap Forward that had taken
place in education by impressive lists of statistics.
Numbers of students engaged in higher education had
increased by 80 per cent, in secondary education 112 per
cent, and in primary education 43 per cent.[6]

---

[4]Peking Normal University, Spare-time Middle School
Mathematics Editing Section, "Spare-time Mathematics
Teaching Outline," Shuxue Tongbao (October, 1958), p. 23.

[5]J. C. Cheng, "Notes and Comment: Half-Work and
Half-Study in Communist China," Pacific Affairs (June,
1959), 32: 178-193.

[6]Chang-tu Hu, "Recent Trends in Chinese Education,"
International Review of Education (January, 1964), 10:
12-21.

TABLE 13

SPARE-TIME MATHEMATICS PROGRAM FOR TSINCHOW
TEXTILE AND MACHINE MANUFACTURING SCHOOL[7]

	Topic	Lecture Hours	Homework Hours
1.	Preliminary concepts: functions and graphs	4	2
2.	Approximate calculations, use of slide rule	10	2
3.	Work with exponents and roots	14	4
4.	Measurement and similarity of triangles	10	2
5.	Solution of quadratic equations and inequalities	16	2
6.	Properties of polygons and polyhedra	8	2
7.	Introduction to trigonometric functions	16	4
8.	Logarithmic and exponential functions	8	4
9.	Properties of points and lines	10	2
10.	Theory of second degree equations	10	4
11.	Trigonometric formulae and inverse trigonometric functions	10	4
12.	Applications of trigonometric formulae	10	4
13.	Trigonometric series	4	2
14.	Factoring of equations	8	2
15.	Theory of logarithms	10	2
16.	Fundamental formulae involving logarithms and rules of use	8	4
17.	Applications of logarithms	12	2
18.	Introduction to calculus and its applications	8	2
19.	Properties of indefinite integrals	8	4
20.	Definite integrals	8	2

[7]Mathematics Teaching and Improvement Team,
"Spare-time Mathematics Outline for Tsin-chow Textile
and Machine Manufacturing School," Shuxue Tongbao (January, 1959), pp. 26-29.

TABLE 13 (Continued)

Topic	Lecture Hours	Homework Hours
21. Applications of integration	8	2
22. Partial differentiation	6	2
23. Introduction to simple differential equations	4	
Total Hours	210	62

Under the additional demands of productive labor, educational activities of the regular school system became frenzied. Schools constructed their own blast furnaces to aid in the steel production effort. In academic matters, some schools advocated "doing five years work in three years" by skipping sections in their texts or abbreviating course work. The Middle School of Shantung Marine College was credited with "making a mouthful" of mathematics, physics, and chemistry.[8] Still other schools changed from a six-year course of study to five years. Educators, flushed with a new spirit of nationalistic dedication, rewrote texts and study outlines eliminating obvious foreign influence. Researchers began a flood of experimental studies to see how the curricula could be effectively condensed. Many of these studies concerned mathematics content and teaching and reveal existing deficiencies.

---

[8]K. E. Priestly, Education in China (Hong Kong: Dragonfly Books, 1961), p. 35.

Period of Educational Adjustment

In the wake of the 1958 innovations, the Chinese educational scene experienced a period of adjustment in 1959. School enrollments continued to grow, burdening facilities which had already been expanded beyond their effective capacity. In the resulting confusion of the previous year, party cadre had firmly infiltrated educational administration. Although spare-time teacher-training programs had been expanded and "knowledgable" persons recruited as teachers, these measures did little to satisfy the growing demand for instructors. Students and teachers began to falter under the growing requirements and privations of the production-oriented campaign. Educators disagreed among themselves as to the direction mathematics education should take. Some believed that additional branches of mathematics, such as calculus and analytic geometry, should be added to school curricula. A few middle schools, such as those attached to the Peking Normal University, had successfully experimented with the teaching of these production-oriented courses. Other educators preferred to reduce the length of middle schooling to four years and produced appropriate mathematics teaching outlines for the purpose.[9] By removing

---

[9]Mathematics Department, Shanghai Normal University, "Four Year System Middle School Mathematics Teaching Outline," Shuxue Jeaoxue (April, 1959), pp. 6-11.

redundant matter from the mathematics curriculum, it was felt that one year's material could be adequately covered in one semester. Despite some limited attempts to raise the quality of school work, academic standards began to deteriorate. An editorial in the People's Daily of December, 1959, cautioned that secondary education should be "fundamental education" with stress on Chinese, mathematics, physics, chemistry, and a foreign language. As a result of educational disruption caused by the Great Leap Forward, the government of the People's Republic had to re-examine the objectives and methods of its educational system, while simultaneously proclaiming the regime's progress in education during its "ten great years" of existence.

Reforms Intended to Upgrade Mathematics Teaching

The year 1960 saw extensive reforms undertaken in mathematics education in the People's Republic of China. These changes were in a response to (1) a lessening of the quality of school work, and (2) experimental findings concerning mathematics learning and teaching in the schools. Both phenomena were a direct result of the activities of the Great Leap Forward. A January article in Shuxue Tongbao entitled, "Opinions on the Goals and

Mission of Middle School Mathematics,"[10] condemned governmental pressure for orienting mathematics teaching toward
production. Its five authors stated that mathematics
should not be used to teach politics, nor should it
become a mere tool for production. Further, they felt
that if these policies were pursued, the quality of
mathematical knowledge would be lowered, which in turn
would hurt the socialist ideals of the state. In light
of the government's backlash following the criticisms of
the "Hundred Flowers" period, so outspoken a complaint
against official policy must have been warranted by grave
complications in the educational field. These opinions
were in direct opposition to the journal's editorial,
"To Raise the Red Flag of the Guidelines and Educational
Principles of the Party and to Struggle to Alleviate the
quality of Mathematics Education,"[11] which pointed out
the great advances in education made under party leadership. A third article on the topic of reform entitled,
"On Whether it is Necessary to Rewrite the Mathematics
Outlines for Elementary and Middle Schools" by a
Kuang Han, agreed that the Great Leap had resulted

---

[10]Pien Shu-yang, et al., "Opinions on the Goals
and Mission of Middle School Mathematics," Shuxue Tongbao
(January, 1960), pp. 38-40.

[11]Shuxue Tongbao (January, 1960), pp. 2-4.

in upheavals in mathematics education.[12] Kuang offered
specific suggestions intended to increase the quality of
mathematics teaching and learning: the study of analytic
geometry should be reinstituted in the schools; elementary algebra, including work with second degree equations,
should be removed from the senior curriculum and taught
in the lower middle school; and trigonometry should no
longer be taught as a combined subject with geometry,
but by itself.

The Middle School Mathematics Research Group of
Peking Normal University joined the debate by publishing
its suggestions for mathematics revision.[13] It proposed
changes based on curriculum experimentation jointly
undertaken with members of the Institute of Psychology.
In particular, it advised that all arithmetic teaching
matter be removed from the middle school and taught at
the primary level; that the study of plane geometry
should be completed in the lower middle school; and that
work involving complex numbers be eliminated from lower
middle school studies and time spent on geometric constructions reduced. Theory involving second degree

---

[12]Kuang Han, "On Whether it is Necessary to Re-
write the Mathematics Outlines for Elementary and Middle
Schools," Shuxue Tongbao (January, 1960), pp. 32-35.

[13]Middle School Mathematics Teaching and Research
Group, Peking Normal University, "Our Ideas Concerning
Present Middle School Mathematics Teaching," Shuxue
Tongbao (January, 1960), pp. 35-38.

equations was to be completed in lower middle school.
Its recommended approach to teaching was to center around
the concepts of function and limit, with analytic geometry, differential calculus, and simple probability added
to the curriculum. Analysis and probability were cited
as being necessary for the industrial needs of the state.
The group felt its proposals would bring the Chinese
mathematics curriculum closer in content and philosophy
to that of China's socialist brothers, Russia and East
Germany.

Peking Normal University's campaign for reform
continued into the month of March with the publication of
an article entitled "Problems in Current Middle School
Mathematics."[14] The research group authors felt that
mathematics teaching was antiquated, that it better
reflected the needs of the seventeenth century than those
of the twentieth. Several specific examples of poor
teaching practice were cited: methods of computation
were unnecessarily involved--series approximations could
be replaced by integration, and difficult problems in
arithmetic solved more easily by algebraic methods; the
teaching of the area of a circle took eleven hours of
instruction and ten thousand characters of text; lessons

---

[14]Mathematics Education Research Group, Peking
Normal University, "Problems in Current Middle School
Mathematics," Shuxue Tongbao (March, 1960), pp. 30-32.

on the volume of prisms, cones, and frustra took thirty-
two hours and sixty thousand characters; the whole of
first-year middle school geometry study was unnecessary--
the material was of such a nature that students would
learn it by themselves from everyday experience. Elabo-
rate ruler and compass constructions were time-consuming
and did little to improve student ability in spatial perception. The study of solid geometry should be replaced
by mechanical drawing, analytic geometry, and some work
in projective geometry. The lengthy discussion, forty-
six hours and twenty thousand characters, on the solution
of second degree equations should be replaced by a consideration of the properties of the discriminate and
supplemented by practice in curve sketching. General
conclusions for improvement urged an algebraic approach
to teaching centering on the concept of function, around
which the various mathematical disciplines could be
coordinated. This proposed approach to mathematics
teaching was a unified one. Joining in the debate, but
with a slightly different purpose for reform, that of
better correlating middle school mathematics with university needs, was a student research group from Peking
Normal University.[15] Students influenced by the

---

[15]Science Research Group, Freshman Class, Peking
Normal University, "Our Views on the Problem of Dovetail-
ing Mathematics Courses in Middle Schools with that in
Colleges," Shuxue Tongbao (March, 1960), pp. 35-37.

educational abbreviations of the Great Leap were experiencing academic difficulties upon entering institutions of higher learning. The research group found that only 72 per cent of college freshmen could keep up with their studies in analysis. They suggested that middle school mathematics instruction be more intuitive in nature, building upon the concepts of function and limit, and that they stress an inductive approach to mathematical thinking. They further urged that the concept of function be studied from several viewpoints, as a physical concept (relationship of data), as a mathematical formula, and as a geometrical relationship--no mention was made of a set-theoretic approach to functions!

From February 24 to March 4, 128 representatives of the Chinese Mathematics Society met in Shanghai to discuss the future directions of mathematics teaching. Two Soviet mathematicians were in attendance as advisors. Those present agreed that mathematics teaching should stress the continuity within the discipline and develop the interrelationship between all branches of mathematics. The delegates further concurred with the recommendations of Peking Normal University that analytic geometry, analysis, and probability and statistics be added to the middle school curriculum.[16] Similar conclusions were

---

[16]Huang Kuo-pao, "The Second Congress of the Chinese Mathematics Society," Shuxue Tongbao (April, 1960), pp. 2-4.

reached at a symposium on the Revision of Mathematics Curricular Materials held later the same month by the Institutes of Mathematics and Psychology of Academica Sinica. The symposium's participants agreed with previous recommendations concerning the modernization and strengthening of the middle school mathematics program, and went on to urge that teaching provisions be made to allow for work in computer science.

The official call for national educational reforms came at the April meeting of the People's Congress, the primary legislative body in the People's Republic of China. Speeches by Yang Hsui-feng, Minister of Education, and Lu Ting-yi, Vice Premier of the state and party spokesman, outlined the new directions education would take. Paramount in the new provisions was the shortening of the twelve-year school cycle to a ten-year one. Economic considerations were the motivating force behind this decision.

> Why do we advocate "approximately 10 years?"
> Because it takes approximately 10 years for
> children who start schooling at six or seven
> years of age to grow to the age of 16 or 17,
> when they will be considered as full manpower
> units.[17]

---

[17]Yang Hsui-feng, "To Undertake Pedagogic Reform Positively and to Develop Educational Enterprise More Abundantly, Quickly, Better and Economically," quoted in Robert Barendsen, "Planned Reforms in the Primary and Secondary School System in Communist China," Education Around the World (Washington, D. C.: U. S. Department of Health, Education and Welfare, 1960), p. 6.

Concentration on production-related knowledge was advocated. Although the abbreviation of the period for schooling required the condensation or elimination of certain school subjects, mathematics teaching was not adversely affected. Indeed, the maintenance of quality mathematics instruction was given priority in the proposed program.

> If language and mathematics are properly mastered, it becomes relatively easy to master physics, chemistry, biology, history and geography. The joint efforts of all teachers are needed in enabling the students to master languages and mathematics properly . . .[18]

In his speech entitled "Teaching Must be Reformed," Vice-Premier Lu especially singled out mathematics teaching for comment. Lu cited the experimental work of the previous two years on mathematics teaching as evidence that much mathematics content could be downgraded and taught at a lower level. Thirteen provinces and municipalities had carried out projects on teaching mathematics in kindergarten. It was found that kindergarten children in their play could learn mathematical operations and facts for numbers not exceeding twenty. Other experiments performed by the Institute of Psychology proved that algebraic techniques could be successfully employed

---

[18]Lu Ting-yi, "Education Must be Reformed," Current Background, No. 630 (Hong Kong: United States Consulate General, August, 1960).

in the fifth grade of primary school. Lu condemned the
traditional redundancy in primary instruction of teaching
the elementary arithmetic operations in seven stages,
i.e., for numbers up to 10, 20, 100, 1000, 10,000, and
1,000,000. He concluded that "all these experiments and
views indicate that the standards of the middle school
graduates under the new system would be raised to the
level of present college freshmen." Lu advocated
adherence to the specific reforms proposed by Peking
Normal University and the Mathematical Society Conference.

Yang Hsui-feng's speech, "To Undertake Pedagogic
Reform Positively and to Develop Educational Enterprise
More Abundantly, Quickly, Better and Economically," was
more caustic in its denunciation of mathematics teaching
and quality:

> . . . much of the mathematics, physics, and
> chemistry now taught in middle schools, in
> particular, is old stuff from the 19th century
> which in no way represents the science and
> technology of today.[19]

He suggested that mathematics teaching be begun at an
earlier age and pushed forward at an accelerated rate.
Arithmetic and elementary algebra teaching of the lower
middle school could be done at the primary level, and
basic college mathematics, including calculus, could be
covered in senior middle school.

---

[19]Barendsen, Education Around the World, op. cit.,
p. 7.

A third speech by Yeh Shêng-t́ao, "Textbooks of Secondary and Primary School Must be Innovated," reinforced Yang's and Lu's positions. Yeh maintained that much of the material in primary mathematics books was repetitious and presented at a level below student ability. Through proper textbook writing and teaching, he believed that the mathematics material previously covered in seven years could be learned in four. He proposed a revolution in teaching that would destroy the old framework. In particular, he declared that the Euclidean system of geometry was antiquated and had to be replaced by more relevant material.[20] All the speakers urged either direct or indirect adoption of the Peking Normal University plan of reform.

Peking Normal University's Program

Acknowledged as the vanguard of reform, Peking Normal University further clarified its proposals by publishing several mathematics teaching outlines designed to advance and accelerate the state's industrial capability. It declared that by proper teaching, mathematics material that encompassed 2,394 hours of instructional time could be covered in 1,944 hours. Their nine-year integrated system, upon completion, would supply students

---

[20]Union Research Service Notes, April–June 1960, Vol. 19 (Hong Kong: Union Research Service, 1960), p. 229.

with a mathematical background equivalent to college
freshmen. The mathematics sequence is as follows:

General Headings	Specific Disciplines	Grades
Simple Algebra	Arithmetic, Elementary Algebra, Geometry	1 - 6
Theory of Elementary Functions	Algebra, Trigonometry Analytic Geometry	6 - 8
Analysis	Differential and Integral Calculus, Ordinary Differential Equations, Probability and Statistics	9
Drawing	Plane Geometry, Solid Geometry, Projective Geometry	7 - 9[21]

An accelerated six-year middle school outline was also
devised. Upon completion of required studies, a graduate
would have the background of a college junior.

General Heading	Specific Discipline	Grades
Algebra	Simple Algebra and Theory of Elementary Functions, Drawing and Surveying (From the nine-year program general headings)	7 - 9
Analysis	Calculus, Ordinary Differential Equations, Vector Algebra, Introduction to Complex Variables, Partial Differential Equations	10-12

---

[21]Middle and Primary School Mathematics Pedagogical Reform Research Group, Peking Normal University, "Suggestions on the Modernization of the Mathematics Curricular Materials for Middle and Primary Schools," Shuxue Tongbao (April, 1960), pp. 4-10.

General Heading	Specific Discipline	Grades
Statistical Mathematics	Linear Algebra, Probability and Statistics	10 - 12

Two technical school mathematics outlines were also put forward. Both courses of study were intended to produce graduates of junior college calibre within a limited time period. (See Table 14)

TABLE 14

PEKING NORMAL UNIVERSITY'S TECHNICAL
MATHEMATICS PROGRAMS

Five Year System for Electricity and
Electronics Technical Schools

General Heading	Specific Discipline
Theory of Functions and Analysis	Advanced Algebra, Plane and Solid Geometry, Fourier Series, Study of Periodic Functions, Series Approximation
Probability and Statistics	Plane and Solid Geometry
Drawing	Projective Geometry

Three Year System for Metallury
Technology

General Heading	Specific Discipline
Theory of Elementary Functions	Advanced Algebra, Trigonometry, Analytic Geometry, Calculus Ordinary Differential Equations

Seemingly in competition with the Peking programs, the Kirin Normal University published its own proposal for mathematics teaching based on experiences acquired during the Great Leap Forward.[22] Kirin's suggestions evolved from extensive practical production experience obtained by visiting one hundred local industries. Teams of students from the university had solved problems concerning the stresses produced in the turbines of the Yangtze Gorge, in the Lung-wang reservoir, and in problems concerning auto production. The findings of one hundred symposia, debates, and study sessions went into producing a mathematics outline for a nine-year school system. The highlight of the Kirin program was a proposal to combine analytic geometry, linear algebra, and advanced algebra into one course within which computational methods would be stressed. In general Kirin's and Peking's programs were almost the same, with the former giving more emphasis to industrial applications of mathematics.

The Climate of Reform

The following months saw continued discussion in Chinese newspapers and educational journals on the need for reform. In June, Yang Hsui-feng gave a second speech on educational reform in which he severely criticized

---

[22]"Kirin Normal University Reforms Mathematics and Education Systems," Kuang-ming Jih-pao, Peking (April 27, 1960), p. 2; translated in Joint Publications Research Service, No. 15, 515.

"the bourgeois ideas of teaching according to ability."

Yang pointed out that bourgeois educationists were

strongly opposed to Peking University's new mathematics

outline.[23] _Shuxue Tongbao_ became a central forum for

the exchange of opinions on mathematics reforms. Among

the many educators who wrote in praise of the proposals,

a majority especially agreed with the introduction of the

study of probability and statistics into the middle

school curriculum. Professor Lin Ch'eng-chu of the

Mathematical Dynamics Department at Peking University

sounded a very progressive note by urging that computer

science and linear programming techniques should also be

taught in middle school, due to their numerous industrial

applications.[24] Passion for reform was evidenced in the

words of the state's spokesman. Initiation of these

reforms, however, would rest in the hands of the mass of

teachers already exhausted by the previous reforms and

developments of the Great Leap Forward.

Period of Retrenchment: 1961-1963

By 1961, the consequences of the Great Leap were

being more fully realized: the expected increase in

---

[23]Fraser, _op. cit._, pp. 365-375.

[24]Lin Ch'eng-chu, "My Views on Middle School Pedagogical Reform," _Shuxue Tongbao_ (May, 1960), pp. 17 and 41.

production had not materialized, agricultural output had been sorely reduced, and the educational system badly demoralized. Educators still produced students who were both "red and expert," but they now stressed "expertness" over "redness." Productive work required of students was lessened, and teachers were allowed more time for lesson preparation. Teaching research groups met more frequently to assist their members with class preparation. Classroom instruction stressed the teaching of fundamentals. Rationalizations for the new policies appeared in the press, explaining that production knowledge requires the acquisition first of theoretical knowledge.

Curricula and quality of instruction in spare-time educational institutions came under attack. Especially singled out for criticism was the agricultural middle school:

> . . . failure of a student to acquire the same basic knowledge [about main subjects like Chinese language and mathematics] as from an ordinary middle school would naturally make a mockery of an agricultural school.[25]

The number of spare-time institutions was reduced, and the remaining attempted to upgrade their offerings. In June, Lin Feng, a member of the Central Committee and

---

[25] Kuang-ming Jih-pao, March 12, 1962, in Surveys of China Mainland Press, No. 2710 (Hong Kong: United States Consulate General, April 2, 1962), p. 15.

educational reformer, condemned the quality of academic
work done during the Great Leap and urged that school
studies be concentrated and standards of quality
stressed.[26] Lin advocated the adoption of the pre-1958
curriculum of the middle school of Peking Normal University by all Chinese middle schools. On July 31, the Ministry of Education issued a "Circular Concerning Implementation of the Teaching Plan for Regular and Primary
Schools." This plan appeared to be based on pre-1958
policies (see Table 15). By September, schools started
using new texts that, while production oriented, reflected
the contents of subject outlines prior to the Great Leap.

Mathematics, A Priority Subject

Mathematics was still a subject of prime importance! A major article in a Spring issue of Hung Ch'i
entitled, "The Function of the Mathematical Method in the
Cognition of the Objective World," concerns the dialectic
nature of mathematics. The author, Ho Tso-Ch'wang, by
using quotes from Mao, Marx, and Engles, attempts to
establish the place of mathematics in the building of a
socialist society. In promoting the study of mathematics
as a priority for socialist youth, Ho writes of a new Leap
Forward in the application of mathematical knowledge:

[26]"Chronology of the Two Road Struggle," op. cit.,
p. 41.

> A new leap forward in this area [mathematics] is
> expected soon. In either the science or educa-
> tion of our country, we must properly estimate
> this tendency of scientific development. Espe-
> cially in the schools, it is important to raise
> the training and mastery of the mathematical
> skills of the students. This training concerns
> not only the study of theoretical sciences, but
> also the problem of the student's ability in
> logical thinking. We must use mathematical tools
> extensively in education, to strengthen the
> student's basic mathematical training. Their
> knowledge of mathematics must be relatively
> strict and profound in order to proceed with new
> creations upon this foundation so as to raise the
> scientific level of our country to an even higher
> standard.[27]

The new middle school curriculum reflected some of the

proposed reforms (see Table 16). Arithmetic was no longer

a subject of the middle school, and some schools began

teaching analytic geometry to their seniors. Work in

classical plane geometry still comprised a major portion

of the mathematics program, in opposition to the Peking

Normal University's recommendations. Now about one-sixth

of class time was devoted to the study of mathematics.

Many teachers found it difficult to adapt to the new

changes. Some complained of the middle school student's

poor ability in arithmetic, indicating their preference

for the retention of arithmetic studies in the middle

school. General student performance in algebra and trig-

onometric studies was considered good in comparison to

their work in geometry.

---

[27]Ho Tso-Ch'uang, "The Function of the Mathematical
Method in the Cognition of the Objective World," Hung-
Ch'i (Red Flag), Peking (May 16, 1962).

TABLE 15

RURAL PRIMARY SCHOOL CURRICULUM
1962 (HOURS/WEEK)[28]

Subject Year	1	2	3	4	5	6
Mathematics	6	6	6	5	6	6
Abacus Practice	-	-	-	2	-	-
Chinese	14	14	14	14	12	12
Drawing	2	2	1	1	1	1
Sports	2	2	2	2	2	2
Music	2	2	2	-	1	1
Productive Labor	-	-	-	1½	1½	1½
Nature Study	-	-	-	-	1	1
Geography	-	-	-	-	1	-
Agriculture	-	-	-	-	2	1
History	-	-	-	-	-	2
Private Study	16	16	17	17	15	15*
Totals	42	42	42	42½	40½	40½*

* estimated

"Little Treasure Pagoda" Schools

To meet the needs of increased "expertness," the
Ministry of Education promoted a dual system of education.
In 1959, Lu Ting-yi devised a plan to establish elite
education institutions with superior facilities and
skilled teachers so that academically talented students
could be nurtured in their development. The year 1961
found the scheme being discussed as a general phase of
educational revision. Selected institutions would become
"key" schools and provide the People's Republic with a
"little treasure pagoda" of scholarly talent. By

---

[28]Compiled from Jan Myrdal, Report From a Chinese
Village (New York: Pantheon Books, 1965), pp. 297-300.

December 1962, the plans for the system became formalized, and standards for selection and operation of "little treasure pagoda" schools were laid down. Two hundred thirty-five middle schools, together with approximately fifteen hundred primary schools and several universities, were selected to participate in the scheme. From this number, thirty-six middle schools and one hundred sixty-two primary schools were singled out for even more exceptional attention.[29] Selected schools were allowed larger operating budgets, reduced labor quotas, and a longer school year. In turn, the academic year of the remaining schools was reduced by one month, and the teaching of foreign language removed from their curricula. By these actions, it was clear that a special path to higher education was being established through the pagoda system, as students without foreign language proficiency could not enter a university. Now, mathematical talent discovered through the national competition could be developed in a pagoda school.

---

[29]Revolutionary Committee, Shantung Provincial Department of Education, "Demolish the 'Little Treasure Pagoda' System of Revisionist Education," Jen-min Jih-pao (December 17, 1967; Translated S.C.M.P., No. 4100).

TABLE 16

GENERAL MIDDLE SCHOOL CURRICULUM CANTON
CIRCA 1962 (HOURS/WEEK)[30]

Subject Year	Junior Middle School			Senior Middle School		
	1	2	3	4	5	6
Language and Litera-ture	6	6	6	6	6	6
Foreign Language	4	4	4	4	4	4
Political Science	3	3	3	3	3	3
History	–	–	3	3	3	–
Geography	3	3	3	3	3	–
Biology	3	–	–	–	–	–
Physics	–	3	–	3	3	3
Chemistry	–	–	3	3	3	3
Essentials of Agriculture	–	–	2	–	–	–
Hygiene	–	–	2	–	–	2
Algebra	5	2	3	3	3	3
Plane Geometry	–	3	2	–	–	–
Solid Geometry	–	–	–	3	–	–
Analytic Geometry	–	–	–	–	–	6
Plane Trigonometry	–	–	–	–	3	–
Drawing	1	1	–	–	–	–
Music	1	1	–	–	–	–
Physical Education	2	2	2	2	2	2
Total periods per week	31	31	33	33	33	29

Gradually, the Communist Party attempted to reas-
sert its authority in the educational field and force a
return to ideological training as a focus for education.
Mao, in a directive of February 13, 1962, urged that the
then present school and examination system be revised.
He stressed the necessity of eliminating foreign and

[30]Chiu-sam Tsang, Society, Schools, and Progress
in China (New York: Pergamon Press, 1968), p. 183.

native dogmas, reducing the student's work load, and the easing of examination standards.[31] Following the Central Committee Plenum in September, official pressure for the re-emphasis of political indoctrination in education increased. By the Spring of 1963, "redness" was again the central theme of Chinese education. Productive labor was once more used as a means of increasing class consciousness, and a new official encouragement of peasant-worker spare-time education fostered proletarianism.

Period of Struggle and Rededication Under the Influence of the Great Cultural Revolution: 1963-

Education now underwent a period of conflict and strife caused by the interaction of two divergent concepts: education as an elitist activity preparing students for expertness in a chosen field, and education as a political vehicle promoting proletarianism among China's masses. School year 1963 saw the official adoption of analytic geometry into the middle school curriculum. In key middle schools, the study of science and mathematics accounted for about 50 per cent of the student's time.[32] Teaching emphasis was still on

---

[31]"Chronology of the Two Road Struggle," op. cit., p. 47.

[32]C. H. G. Oldham, "Science and Education," Bulletin of Atomic Scientists (June, 1966), p. 43.

production-oriented applications of mathematics, but now
the necessary theoretical background was better taught.
Teachers attempted to isolate and reduce student mathematical deficiencies. In particular it was noted that
students incurred difficulty in associating mathematical
concepts with problem situations and had little understanding of the theory and use of polar coordinates in
analytic geometry.[33] Criticism was once again directed
at the teaching of geometry in the middle school. It was
felt that the subject neither developed the students'
concept of space nor trained them in the use of deductive
reasoning based on an axiomatic system. Geometry was
still taught by rote and rule, encouraging unthinking
memorization.

Stress on Spare-Time Education

In a speech early in 1964, Mao, for the second
time in a year, urged educational revision.[34] He advocated that:

---

[33]Li Hsi-yen, Hopei Normal University, "On Some
Problems Associated With the Teaching of Polar Coordinates in Middle School Geometry," Shuxue Tongbao (June,
1964, pp. 15-20; Ku Ch'ing, "The Problems Existing Among
Students as Found From a Test and How to Improve the
Quality of Mathematics Teaching," Shuxue Tongbao (March,
1964), pp. 12-15.

[34]Mao Tsê-tung, "Comment on 'Views Advanced by a
Middle School Principal Concerning the Question of Alleviating the Work Load of Middle School Students'," (March
10, 1964; Current Background, No. 888).

1. The total period of schooling be shortened.

2. Curriculum could be reduced by half.

3. The stress on examinations should be discontinued.

Mao further cautioned that academic demands were becoming overburdening for students. These comments were especially directed at the "golden pagoda" system of schools. On March 10, he again openly attacked school curricula and the examination system. A new era of party-promoted spare-time education was instituted. Half-work, half-study programs that had previously been used only in rural areas during the Great Leap Forward were now instituted in the cities. Increased numbers of spare-time and part-time schools sprang up around the country. Their more organized curricula reflected experience gained in the 1958-60 period and attempted to better provide for local needs. A half-work, half-study mathematics curriculum for technical middle schools as devised by a school in Tientsin is given in Table 17. Its program included the study of calculus and appears quite complete.[35]

---

[35]First Part-work Part-study Middle Technical School of the First Bureau of Machine-building Industry of Tientsin Municipality, "Mathematics Curricular Program for Part-work Part-study Middle Technical Schools," Shuxue Tongbao (September, 1965), pp. 2-6.

TABLE 17

HALF-WORK HALF-STUDY MATHEMATICS PROGRAM,
TIENTSIN TECHNICAL MIDDLE SCHOOLS[36]

Subject	Periods Expended		
	Total Time	Classroom Instruction	Homework
Algebra	82	70	12
1. Simple functions and their graphs	12	10	2
2. Work with approximations	10	9	1
3. Quadratic equations and graphs	8	6	2
4. Simultaneous solutions for two second degree equations	6	4	2
5. Introduction to series	6	5	1
6. Work with powers and roots	10	8	2
7. Theory and application of logarithms	18	16	2
8. Slide rule practice	12	12	-
Trigonometry	62	52	10
1. Properties of angles and triangles	8	6	2
2. Measurement of angles	4	4	-
3. Trigonometric functions of any angle	8	8	-
4. Graphs of trigonometric functions	12	10	2
5. Use of trigonometric functions in volving triangle problems (sum and difference formulae)	8	6	2
6. Properties of the sine curve	12	10	2
7. Inverse trigonometric functions	10	8	2
Solid Geometry	14	10	4
1. Various polyhedra and their problems	6	4	2

_____

[36]Ibid., p. 4.

TABLE 17 (Continued)

Subject	Periods Expended		
	Total Time	Classroom Instruction	Homework
2. Surface area of a volume of revolution	6	4	2
3. Properties of a sphere	2	2	–
Advanced Mathematics	92	79	13
Plane Analytic Geometry	24	20	4
1. Use of coordinates	2	2	–
2. Properties and Graphing of a straight line	8	6	2
3. Second degree curves	14	12	2
Elementary Calculus	42	37	5
1. Concept of limit	8	7	1
2. Functions and continuity	4	4	–
3. Definition of derivative	16	14	2
4. Application of derivatives	10	8	2
5. Indefinite integral	12	10	2
6. Definite integral	6	6	–
7. Applications of integration	8	6	2
Extra Curricular Activities	30		30
Total periods spent in mathematics studies	280	211	69

Educational Conflict

Despite constant verbal harangues and pressures from the Communist Party, educators still strove to improve the quality of the middle school curriculum. In about 1965, the study of calculus was added to the mathematics program. Two periods a week in the last semester of senior middle school were devoted to an introductory study of differential and integral calculus. Emphasis was on the use of the calculus as a mathematical tool, rather than on a theoretical study of infinite approximation techniques. Under party insistence, a strong emphasis on productive work had been reinstituted in the schools. Between the rigors of the curriculum and the physical demands of labor, many students and teachers functioned in a state of near exhaustion. Pressure was placed on schools to accept students from worker-peasant background on social-class merit alone. Most of these students found it impossible to satisfy the expectations of their teachers. Discipline and the academic climate of the schools began to deteriorate. Students from worker-peasant backgrounds complained that teachers openly discriminated against them. In 1965, teachers from Shanghai schools held debates on "What Type of Students Should the Teacher Prefer?"[37] The party-guided consensus of opinion ruled that they should intensify their viewpoints in favor of students with proletarian backgrounds. By the beginning of 1966, the tenor of the articles in Shuxue Tongbao began to reflect this concern of teachers towards the deterioration of mathematics

---

[37] Union Research Service, Vol. 47, No. 24 (Hong Kong: Union Research Institute, 1967), pp. 337-352.

instruction. Several expositions offered solutions to motivational problems involving mathematics learning.[38] In this same period, the commentaries in all Chinese media began to reflect the national vogue of "how studying the works of Chairman Mao helped me to . . ." Typical of this journalism, an article entitled "My Experience in Improving Mathematics Pedagogy According to Chairman Mao's Ideology," lists the author's Mao-inspired techniques for successful teaching:

1. Attempt to solve major contradictions first.
2. Draw the students' attention to the internal relation between things in the same field of knowledge.
3. Teach students to master principles and important theories before going into details.
4. Employ ideological education to promote the learning initiative of the students.[39]

Apparently, under the influence of Mao thought, many mathematics teachers re-examined their classroom conduct and attempted to remedy their defects.

The Cultural Revolution

Mao continued making major pronouncements concerning

---

[38]Chou Hsueh-Ch'i, "Some Basic Appreciation as to Developing Student's Ability to Relate Theory to Practice in Mathematical Thinking," Shuxue Tongbao (January, 1966), pp. 3-7; K'ang Chieh Middle School, "Some Experimentation in Mathematics Teaching Regarding Student Motivation Towards a Positive and Active Attitude," pp. 8-10.

[39]Wu Hsueh-lu, "My Experience in Improving Mathematics Pedagogy According to Chairman Mao's Ideology," Shuxue Tongbao (February, 1966), pp. 4-6.

a restructuring of the educational system. In a speech delivered at the Hangchow Conference in December of 1965, he reveals a glimpse of his epistomological philosophy that is strikingly similar to that held by some contemporary Western proponents of educational revolution:

> I am rather skeptical about the present educational system. The whole course of education from primary school to university takes in all sixteen or seventeen years, up to more than twenty years. During the period, one sees not how cereal plants are raised, how the workers work, how the peasants till land, how commodities are exchanged, and becomes poorer and poorer in health. This is truly harmful. I have told my child: "Go to the countryside and tell the poor and lower-middle peasants this. Tell them that your father says one gets more stupid after going to school for more than ten years. Ask your uncles, sisters and brothers to be your teachers and learn from them." As a matter of fact, a child gets in touch with a lot of things when he is between one year and seven years old before he goes to school. He learns to talk when he is two years old, and when he is three, he quarrels with people. When he grows older, he digs soil with small tools and mimics the adult in labor. This is observing the world. The child has learned some concepts. The dog is a general concept, while the black dog, the brown dog are particular ones. The brown dog in his home is concrete. The notion of man has done away with many things . . . the difference between man and woman, the difference between the adults and children, the difference between Chinese people and foreigners . . . and only the characteristics which differentiate man from other animals are left. Who has seen "man"? Only Tom, Dick and Harry can be seen. The concept of "house" is also seen by nobody, and only the concrete houses, the European-style houses of Tientsin, the houses built around courtyards in Peking are seen.[40]

---

[40]Mao Tsê-tung, "Speech Delivered at Hangchow Conference," December 21, 1965; Current Background, No. 888, op. cit., p. 16.

In a published letter to Lin Piao on May 7, 1966, Mao provided the final impetus to the radical reformation of education thinking. In writing on the needed changes in Chinese society, he assigns a specific role for students:

> This holds for students too. While their main task is to study, they should in addition to their studies, learn other things, that is, industrial work, farming and military affairs. They should also criticize the bourgeoisie. The school term should be shortened, education should be revolutionized, and the domination of our schools by bourgeois intellectuals should not be allowed to continue.[41]

On May 25, disruptions broke out at Peking University and soon spread to many other Chinese schools. A period of intense ideological and intellectual reform had begun-- the Great Cultural Revolution. This movement advanced from the schools to factories and finally the countryside. Soon the whole nation was engulfed in the growing wave of reform championed by a group of disenchanted students who eventually became known as the "Red Guard." An editorial entitled "Sweep Away All Monsters," stating the purposes of the Cultural Revolution, appeared in the June 1 edition of the People's Daily and simultaneously in many periodicals including Shuxue Tongbao and Acta Mathematica Sinica. It called upon the workers, peasants, and revolutionary cadre and intellectuals to "smash the shackles imposed on their minds by the exploiting classes"

---

[41]Current Background, No. 888, op. cit., p. 17.

and to "route the bourgeois 'specialists,' 'scholars,' 'authorities,' and 'venerable masters'." These state-provided condemnations supplied license for open student attacks on teachers. Many experienced and skilled teachers who had received Western training were singled out as bourgeois revisionists, publicly humiliated, and forced to confess to their crimes. The actual numbers of such teachers were indeed great, as a survey of the time revealed that 98 per cent of the faculty in forty-six institutions of higher learning had received their academic training in pre-liberation China and therefore were contaminated by "the class spirit of the bourgeoisie."[42] Complete disruption of formal education forced schools to close in the summer of 1966. The Communist Party had emerged victorious in its struggle with bourgeoisie educators to secure education as a political entity. Party reforms for education were listed in a decision of the Central Committee on August 8. Besides liberating education from the influence of bourgeois intellectuals, the reforms included replacing scholasticism by the methods of materialistic dialectics, and nurturing a socialist consciousness in students by emphasizing pro-letarian politics and productive labor.

---

[42]Hwa Yu, "China," Christian Century (September 30, 1970), p. 1170.

194

Attempts to Restore Order in the Schools

Spring of 1967 saw repeated public calls by the
Central Committee and the State Council for students to
return to their classes. A March 7 editorial in the
People's Daily entitled "Primary and Middle Schools
Resume Classes to Make Revolution," suggested students
return to class and continue their revolutionary activities within the school system. These pleas went unheeded.
The People's Liberation Army (P.L.A.) was called upon to
help restore order in the schools. By the summer, units
of the P.L.A. had established themselves in some schools
and were attempting to institute a semblance of educational activities. The army's limited experience in
formal education resulted in schools being organized on
a military basis, with the students grouped into squads
and companies. Lessons centered around the study of the
thoughts of Chairman Mao and military drills. Even with
the army occupying many schools, the conflict between
teachers and students and among Red Guard factions continued, preventing a return to normalcy. Finally in
March of 1968, Mao issued a directive ordering the army
to take firm action in re-establishing the educational
functions of the state.[43] Army units were to restore
some semblance of discipline and prepare the way for

[43]Current Background, No. 888, op. cit., p. 25.

"work-propaganda teams." Teams composed of factory or
agricultural workers entered schools, forming alliances
with revolutionary students, teachers, or cadre, and led
the struggle to reform education. These three-in-one
alliances, workers, student-teachers, and cadre, insured
that proletarian and political considerations would be
given due attention in the revised school materials. At
first, their activities were limited solely to propaganda--
conducting Mao-thought classes and self-criticism sessions
among students and school personnel; but gradually their
activities moved towards academic concerns. Educational
experiments similar to the Yenan period and the Great
Leap Forward now took place in every school. All education was now basically proletarian in nature, and as
such reflected immediate social priorities. Curricula
were abbreviated, "barefoot" teachers, lacking academic
credentials but victims of the class struggle, replaced
regular classroom teachers. Factories and communes were
given direct responsibility for running schools. Productive labor requirements and military training were
increased. Mao thought became the focus of all learning
activities. Teaching-research groups were disbanded and
principals and department heads removed from authority,
turning educational institutions into classless societies.
The Ministry of Education, reduced to a nominal existence
at the beginning of the Cultural Revolution, now lost all

vestiges of power. Examinations, including those for entrance to higher institutions of learning, were abolished and replaced by party recommendations based on class background, work experience, and service to the Communist cause.

The Lishu Program

Although programs in different schools varied widely, they were united in their attempted adherence to Mao's pronouncement of May 7, 1966. The May 12, 1969, issue of the People's Daily carried a draft proposal by a revolutionary committee of Lushu County, Kirin Province, for a new primary and middle school program.[44] The program received national attention and became the center of a debate on the exact direction educational reform should take. Even though the Lishu plan was designed for rural education, it amalgamated the prevailing practices and obviously appealed to the Communist Party, as testified to by its exposure in the People's Daily. Some of the main points of this program were:

1. Schools would be run by labor brigades, communes, factories, and local party branches. Revolutionary committees composed mainly of poor worker-peasants

---

[44]"Draft Program for Primary and Middle Schools in Chinese Countryside" (Peking: May 13, 1969; in S.C.M.P. 4418, May 19, 1969), pp. 9-15.

would be given the primary responsibility of administering the schools.

2. Mao thought would be the guiding force for all activities and would be employed to produce a political consciousness cognizant of the class struggle. Emulation of the People's Liberation Army would be attained.

3. Teachers would base their instruction on a proper political attitude, instilled by rectification if necessary. Schools would become socially classless, exemplified by close student-teacher relationships. Expansion of extra-curricular activities would link the family, school, and society.

4. In schooling, a continuous nine-year system would be followed--five years primary and four years middle school. The school year would contain 35-40 weeks.

5. Teaching methods would be concise and stress practical applications of the knowledge obtained. Curriculum would be designed to meet local needs. The primary program of study would include five subjects, politics and language, arithmetic, revolutionary literature and art, military and physical education, and productive labor. A similar program would be followed in the middle school: education in Mao Tse-tung thought (including studies in history), basic knowledge for agriculture (including mathematics, physics, chemistry, and economic geography), revolutionary art and literature

(including language), military training, physical culture, and productive labor.

6. Regular academic studies should account for about 60 per cent of study time in middle school and not less than 70 per cent in primary school. Open book examinations should be used and no grades given.

By the summer, this program had received general acceptance, and with slight modifications became the plan for the new school system. Without the central coordination of the Ministry of Education's subject outlines, curriculum content emphasis varied from school to school, according to local needs and competences. Major textbook publishing houses, such as the People's Press, ceased producing standardized school books. Revolutionary committees were called upon to compile teaching materials that would adequately reflect local conditions and needs.[45] Experienced teachers still caught up in the trauma of the assaults and humiliations of the previous two years were reluctant to assist in the re-editing of texts lest they again be accused of promoting a "bourgeois world outlook." The ivory tower system of key schools

---

[45]China News Agency, "Middle School Teaching Material Compiled Under Working Class Leadership," Shenyang, (March 18, 1969); in S.C.M.P. 4383 (March 25, 1969).

was absorbed into the new proletarian system. Teacher research and study groups for curriculum improvement were disbanded and replaced by politically inclined committees. Mathematics competitions were halted.

Mathematical Reforms Under the Great Cultural Revolution

Specific knowledge concerning the teaching of mathematics in the present educational system is fragmentary due to the flexibility of curriculum design; however, several facts concerning its status can be distilled from available resources:

1. Mathematics is still a priority subject in Chinese schools.

2. The inclusion of political materials into mathematics studies has increased greatly.

3. New Mathematics curricula are abbreviated, narrower in scope, and lower in quality than pre-cultural revolution schedules.

4. Mathematics studies are designed for immediate social needs rather than as a preparation for higher learning.

Popularism

To a great extent, the emphasis in local curricula reflects the biases and experiences of the schools' controlling agents, the revolutionary committees. Thus

mathematical studies in an institution run by a factory
are dominated by production problems. Unfortunately, in
such situations, the academic prejudices of the "barefoot" teachers are detrimental to the student's learning.
In one Peking middle school, factory workers convinced
their charges that requiring studies in geometry and
trigonometry preceding mechanical drawing was a bourgeois
"trick."[46] Revised instruction of the subject relied on
the use of actual machine parts and sample drawings,
rather than theoretical knowledge. In a contrasting
situation, a peasant carpenter promoted the study of
trigonometry in explaining to a class how knowledge of
trigonometric functions could assist in the construction
of a room.[47] It appears that the new focus of mathematics learning centers around accounting principles,
abacus practice, surveying, mechanical drawing, and
production problems. As a result, most mathematics curricula have been drastically reduced. Peking Middle
School Number 23 has combined mathematics, physics, and
chemistry into one production course. A common practice

---

[46]China News Agency, "Peking Middle School Pushes
Forward Educational Revolution," Peking (May 15, 1970);
S.C.M.P. 4663 (May 25, 1970), p. 16.

[47]"Shihchingshan Middle School Establishes the
Idea of 'Study for the Revolutionary Cause'," Jen-min
Jih-pao (May 15, 1960; S.C.M.P. 4423 (May 26, 1969), pp.
6-7.

is to combine the previously separate studies of algebra, geometry, and trigonometry into one general mathematics class. Following these practices one Tai Chow middle school reported the consolidation into one diagram of knowledge on polyhedrons that formerly took 102 pages of instruction. The Red Flag Middle School in Penhsi, Liaoning, has reduced its total number of mathematics lessons from 1,100 to 500. One of Red Flag's innovations was to remove citations of foreign accomplishments from their revised texts.[48] Similar consolidation in teaching materials and practices resulted in a Kirin Middle School reducing its curriculum by two-thirds.

Instruction has become more intuitive, building upon the students' experiences, particularly those involved in productive labor tasks. Mathematics instruction avoids the abstract nature of the discipline and concentrates on concrete applications. An editorial in the April, 1970, issue of Hung Ch'i described an interesting and mathematically quite correct analogy used by a Mao propaganda team at Tsinghua University in teaching calculus:

> In the past, the concepts of differential and integral calculus were derived from piles of axioms and theorems and were very mysterious

[48]"Hung Ch'i Middle School in Penhsi, Liaoning, Forms Textbook Group," Jen-min Jih-pao (May 8, 1969); in S.C.M.P. 4419 (May 20, 1969), pp. 6-7.

and unfathomable. Now the concepts are illustrated by familiar instances in production.
For instance, when a bench worker processes a
metal piece into a round shape with a file,
every single movement forms a short straight
line and finally the lines combined result in
curves. This process of turning a whole into
parts and parts into a whole vividly presents the
concepts of differential and integral calculus.
Worker students commented after studying: "After
all, there's nothing mysterious about calculus.
A mere file pierces the myth. Higher mathematics
comes back into the hands of us laboring people."[49]

Indoctrination

Of all the post-cultural revolution innovations in

mathematics education, the most extensive is an increased

emphasis on political indoctrination as a central part of

the instructional process. A recurrent theme in mathe-

matical computations is the past exploitation of the

working class by capitalist landlords and K.M.T. reac-

tionaries. Typical of this new emphasis is a mathematics

lesson entitled, "How the poor and low middle [class]

peasants were exploited in the old society," taught at a

Kiangsu primary school.[50] Peasant-supplied examples of

the class struggle abound in the new textbooks of the

People's Republic. Chi Fu-ch'ing, a peasant, supplied

---

[49]"Strive to Build a Socialist University of
Science and Engineering," Hung Ch'i (August, 1970).

[50]China News Agency, "Poor and Lower-Middle [Class]
Peasants Compile Teaching Material," Peking (January 17,
1969); in S.C.M.P. 4343 (January 22, 1969), pp. 24-26.

the following questions for the arithmetic text of a

school belonging to the Yueh-chin work brigade:

1. Before the liberation, how many persons were there in Hsi-ts'un-pien who worked as hired hands and how many were sold as child-wives?
2. Poor peasant Chi Fu-ch'ing farmed 15 mou of paddy for the landlord and was required to produce 620 catties of paddy per mou a year. What was the total quantity of paddy Chi-Fu-ch'ing produced a year?
3. The landlord paid Chi Fu-ch'ing only 6.5 piculs of rice [one picul of rice being the equivalent of 200 catties of paddy] in wages a year. How many catties of paddy could Chi Fu-ch'ing have after one year's labor? How many catties of paddy did the landlord expropriate from him? What was the percentage of the total output that was expropriated by the landlord? What was the percentage of the total output that Chi Fu-ch'ing received as wages after toiling for a whole year?[51]

An article entitled "Revolution in Education Brings About

a New Outlook," reveals how such materials are being used

in classroom situations:

The problem was: "When worker Tung was six years old, his family was poverty-stricken and starving. They were compelled to borrow five dou of maize [215 pounds] from a landlord. The wolfish landlord used this chance to demand the usurious compound interest of 50 per cent for three years. Please calculate how much grain the landlord demanded from the Tung family at end of the third year." The students were shocked at the figure they worked out. The Tung family owed the landlord nearly 17 dou of maize. This was usury with a vengeance! How vicious the landlord was! Where could the Tung

[51]"Several Arithmetical Questions," Jen-min Jih-pao (January 14, 1969); S.C.M.P. 4355 (February 7, 1969), p. 12.

family get so much grain to repay such a debt?[52]

The instruction proceeds to explain how the landlord seized the Tung's land and rented it back to them. When the family could not meet the rent payments, they had to give the landlord one of their children. Certainly classroom topics such as these will foster the party's desired socialist consciousness, but probably at the expense of obscuring some of the mathematical concepts involved.

Chinese Mathematics Education in the 1970's

The mathematics education scene in the People's Republic of China in the 1970's is one that is unique in modern Chinese history. It is a product of the needs and aspirations of the Chinese people as interpreted by the Communist Party and arrived at through an evolution and assimilation of experience. By breaking with Western educational traditions, the Chinese are attempting to arrive at an educational model that will prove most efficient for their present stage of development. While the disciplines taught and techniques used will provide a basic education for China's masses and produce literate technicians and knowledgable farmers, the question remains, "Will the Mao-inspired education, with its

---

[52] China News Agency, "Revolution in Education Brings About New Outlook," Peking (February 2, 1969); in S.C.M.P. 4355), p. 17.

limited mathematics exposure, provide the People's Republic with sufficiently skilled workers to achieve the industrial goals it has set for itself?" The answer to this question remains to be found in the performance of the present generation of Chinese students after they enter the labor market.

Conclusions

By 1958, the Communists had met the challenges of preserving and expanding the inherited national school system. In their new found confidence, they rejected available Soviet assistance and sought once again to develop an educational model consistent with worker-peasant needs. Under the demands of the Great Leap Forward education gains were designed to parallel industrial advances. A variety of proletarian-oriented educational experiments were attempted. Spare-time and half-work half-study school programs flourished. Regular school curricula, including mathematics studies, were drastically abbreviated. Extreme productive labor requirements took a toll of student time and energy. These excesses lowered mathematical standards to a level that warranted national concern. Professional groups as well as independent educators urged for a return to a rigorous standardized curriculum. The years between 1960 and 1963 saw a return to the classical modern curriculum in a renewed quest for

academic excellence. Education in China now reached an
elitest extreme with the establishment of a system of
privileged schools. In their drive for academic achievement, some of these schools expended as much as 50 percent
of class time in the study of mathematics and science.

While this spirit of educational rigor was intended
to accommodate the state's scientific goals, the exclu-
siveness of the school system's student selection process
violated some basic socialist tenets. Obviously, there
was not equal educational opportunity for all as children
from worker-peasant backgrounds were discriminated against!
The Communist Party, led by Mao, began open attacks on the
school system, its structure and content. The demands
for radical reforms became more insistent and voracious
until finally in May of 1966, Mao called upon the students
to overthrow the educational system. They responded and
the Great Cultural Revolution swept over China.

After a prolonged closing, schools reopened reveal-
ing a completely restructured system dependent on proletarian relevance. Factories and communes now ran the
schools. Mathematics studies were shortened and directed
towards immediate agrarian or industrial applications.
All study became a collective endeavor and tests were
abolished. With the institution of these reforms, it
appears that Chinese mathematics education has finally rid
itself of the traditional influence that has for so long
limited its effectiveness!

CHAPTER V

AN EXAMINATION OF SELECTED INNOVATIONS
IN CHINESE MATHEMATICS EDUCATION
DURING THE COMMUNIST PERIOD

As in describing so many undertakings of the
Chinese Communists in their reshaping of the lives and
aspirations of the Chinese people, the policy of "walking on two legs" best characterizes their actions concerning mathematics education reforms. Several different
but simultaneous approaches have been employed in improving the general academic mathematical climate in the
People's Republic since 1949. While some practices, such
as the founding of pedagogical universities or the institution of mathematical competitions, were directly influenced by association with the Soviet Union, others, like
the proliferation of popular mathematics booklets or the
combining of political indoctrination with mathematics
instruction, are inherently Chinese. Such diverse facets
of reform taken together reveal a concentrated campaign
to improve mathematics instruction that reflects a
national priority. To fully appreciate how intense this
effort has been, several aspects of this movement are
isolated and examined in depth.

The Training of Mathematics Teachers

Expansion of general educational facilities under the present mainland Chinese government has included concerted efforts to train sufficient numbers of teachers to staff the new schools. For the most part, these efforts have been directed through the inherited system of normal schools. At present, three levels of teacher training institutions exist in Mainland China--the junior normal school for preparing kindergarten teachers, the senior normal school for primary teachers, and the higher normal school or university which provides teachers for middle schools and junior and senior normal schools. Higher normal schools are divided into three categories--normal or pedagogical universities, normal colleges, and normal professional schools. All operate at the post-secondary level. Normal colleges and universities offer four-year courses; these schools differ in their availability of course offerings and organizational complexity. The higher normal professional school trains teachers for the junior middle school through a two-year program. Courses of study range from three to five years in length; short-term programs lasting from six months to two years are given to prepare graduates of junior middle schools as secondary teachers. During its period of educational

rehabilitation and consolidation, the government emulated
a Soviet model by establishing teacher-training institutions independent of parent organizations by merging
separate departments or schools of education of various
universities into single higher normal schools. Attempts
were made to distribute normal schools geographically
over a wide area so that both rural and urban locals
would have access to teacher training facilities.[1] As a
result of these innovations, the number of independent
higher normal institutions in China rose from twelve in
1950 to thirty-one by 1953, and to approximately sixty
by 1960. It would be safe to speculate that during the
years of internal conflict and purge since 1960, this
number has not increased appreciably. These new, more
specialized, normal schools usually limit their instruction for teacher preparation to twelve academic disciplines, of which the priority subjects of chemistry,
physics, and mathematics are always included.[2] Supplementing formal instructional programs are various correspondence, television and radio inservice, and part-
time study programs.

---

[1]For a detailed discussion of the aspect of educational innovation, see: Ying Cheng Kiang, "The Geography
of Higher Education in China" (unpublished Ph.D. thesis,
Columbia University, New York, 1955).

[2]For listings and locations of some of these
schools, see Appendix B.

In satisfying its urgent need for elementary and secondary teachers, the government has recruited staff from several sources outside the supply of trained normal school graduates. These sources have included reoriented and politically acceptable pre-liberation-trained teachers, former military personnel, middle school and university graduates, housewives, "knowledgable" peasants and workers, and party cadre. Despite the increase in numbers of nonprofessional school personnel, resulting from the reforms of the Cultural Revolution, the majority of middle school teachers are the product of normal schools.

Formal Training Programs

Middle school mathematics teachers trained in the normal system must complete work in four areas of involvement--political training, educational studies, mathematical studies, and practice teaching experience. In the higher normal four-year program of 1956, mathematical studies accounted for 53 per cent of school time.[3] The student could select a limited number of electives from among differential geometry, the history of mathematics, foundations of modern algebra, theory of computation,

---

[3]Ministry of Education, People's Republic of China, "Introduction of Teachers College Mathematics Training Program," Chung-hsüeh Shu-hsüeh (November, 1956), pp. 41-45.

astronomy, surveying, statistics, and the Russian language. Beginning in the second year of studies, observation visits to primary and middle schools are undertaken. During their sixth and seventh semesters in school, students spend twelve weeks engaged in student-teaching. At this time they are expected to teach three or four mathematics classes daily. The usual class teacher becomes part of their instructional staff and supplies them with a critique of their classroom performance at the end of each day. Student teachers are further expected to become involved in the extracurricular activities of their charges, assisting them in organizing recreational games, athletics, preparing wall newspapers, and joining regular school officials in presiding at parent meetings. Besides meeting the academic demands of this program, successful graduates must pass final examinations in Marxist-Leninist philosophy, educational science and methods, mathematical analysis and principles, and concepts of elementary mathematics. A detailed outline for a four-year mathematics program is given in Table 18. This program of teacher preparation appears quite complete. Laboratory sessions are devoted to performing required experiments for experimental courses such as physics, constructing concrete models that can be used as teaching aids, and doing assigned mathematics exercises. The study of analysis considers topics that would be found in a good advanced calculus course, including an introduction to ordinary differential equations, and line and contour integration.

TABLE 18

REQUIRED PROGRAM OF STUDY FOR A MATHEMATICS
MAJOR IN A FOUR-YEAR NORMAL COLLEGE (1956)[4]

Required Course Work

Subject	Weekly Hours by Year and Semester								Total Subject Hours
	I		II		III		IV		
	1	2	3	4	5	6	7	8	
History of Communist Party			4	4					140
Political Economy					5	5			145
Marxism-Leninism	4	4							140
Dialectic Materialism							4	5	113
Psychology	2	3							87
Educational Science			3	4					122
History of Education					3	3			87
School Hygiene							2		24
Physical Education	2	2	2	2					140
Russian Language	3	3	3	3					210
General Physics			4	5	5	4			291
Mechanics							5	5	125
Drawing		2	2						70
Analytic Geometry	6	4							176
Modern Geometry					4	4			116
Foundations of Geometry							3	4	88
Advanced Algebra	2	2	3	3					175
Real Number System								4	52
Mathematical Analysis	6	6	6	6					420

---

[4]Ibid., pp. 42-43.

TABLE 18 (Continued)

Required Course Work

Subject	I		II		III		IV		Total Subject Hours
	1	2	3	4	5	6	7	8	
Theory of Complex Variables					3	3			87
Principles and Concepts of Elementary Math									391
1. Plane Geometry	3	3							105
2. Solid Geometry			2	2					70
3. Algebra					5				90
4. Real Number System						6			66
5. Elementary Functions							5		60
Methods of Teaching Math					4	4	4		164
Total Hours	28	29	29	29	29	29	23	18	
Weeks/Semester	18	17	18	17	18	11	12	13	

Schedule of Required Laboratories

Subject	1	2	3	4	5	6	7	8
Physics			2	2	2	1		
Mechanics							2	2
Drawing	1	1						
Analytic Geometry	2	1						
Modern Geometry					1	1		
Advanced Algebra	2/3	2/3	1	1				
Analysis	2	2	1	1				
Principles and Concepts of Elem. Math	1	1	2/3	2/3	2	2	2	
Methods of Teaching				1	1	1		

Programs of study offered in the two-year professional schools parallel those of four-year institutions in
their organization.[5] Four broad areas of involvement are
again required, with mathematical studies accounting for
60 percent of school time. A shorter student teaching
experience of four weeks is undertaken in the fourth
semester. Final examinations follow the same format,
with the exclusion of analysis, as those of four-year
schools.

An example of a five-year program of preparation
for middle school mathematics teachers is provided by the
offerings of the Shanghai Pedagogical University.[6]
Founded in 1951, this school offers specializations in
thirteen subjects, including mathematics. During the
period of educational entrenchment, this university
became a "key" institution in the "golden pagoda" system.
The program described in Table 19 was in effect on the
eve of the Cultural Revolution and represents quality
teacher training in the People's Republic of China.

---

[5]Ministry of Education, People's Republic of
China, "Mathematics Schedule for the Junior Normal College," Chung-hsüeh Shu-hsüeh (November, 1956), pp. 37-
40.

[6]For a detailed discussion of this school and its
program see Frank Swetz, "Training of Mathematics Teachers in the People's Republic of China," American Mathematical Monthly (December, 1970), pp. 1097-1103.

TABLE 19

CURRICULUM FOR MATHEMATICS TEACHERS AT
THE SHANGHAI PEDAGOGICAL UNIVERSITY[7]

Subjects	I	I	II	II	III	III	IV	IV	V	V	Total Subject Hours
(Weekly Hours by Year and Semester)	1	2	3	4	5	6	7	8	9	10	
Political Training											
History of Communist Party	2	2									4
Political Economy			2	2							4
Communist Philosophy					2	2					4
Required Mathematics											
Analysis	6	6									12
Higher Algebra		4									4
Analytic Geometry	4										4
Projective Geometry				2							2
Differential Equations						4					4
Func. of Complex Variables								4			4
Func. of Real Variables							3				3
Probability and Statistics						4					4
Differential Geometry								2			2
Physics				5	5	5					15
Theoretical Mechanics							5				5
Technical Drawing	1	1									2
Teaching Methodology								3			3
Textbook Research							x	x	x	x	
Elective Mathematics											
Theory of Functions											
Modern Algebra											Total
Differential Equations											of 200
Applied Statistics											Hours
Numerical Computations											Per
Applied Analysis											Year
Total Hours	13	13	2	9	7	15	8+	9+			
Weeks Per Semester	18	17	18	17	18	17	18	17			

[7] *Ibid.*, p. 1100.

Although this program exceeds those previously examined in mathematical rigor, less emphasis is given to pedagogical training. Pedagogical instruction consists of one three-hour course (per week). Practice teaching is conducted over a six-week period during the fourth year of studies. Two hundred hours of mathematics electives are required in the fifth year of the program. Concurrently, senior students conduct textbook "research." At the present time, such research consists of incorporating the thoughts and teachings of Chairman Mao into existing mathematics texts. In an earlier period, this effort would have been expended in translating Soviet texts and lesson plans into Chinese.

A comparison as given in Table 20 between the mathematical portions, the Shanghai teacher-training program, and the recommendations of the Committee on the Undergraduate Program in Mathematics (CUPM) for the training of secondary teachers of mathematics may prove informative. For comparative purposes, Chinese class time has been translated into semester hours (one hour/week for a semester = one semester hour). A semester hour would roughly approximate a credit hour in an American school; therefore, an American course of three credits would correspond time-wise to a Chinese course of three semester hours.

TABLE 20

COMPARISON OF MATHEMATICAL REQUIREMENTS FOR
MATHEMATICS TEACHER TRAINING IN MAINLAND
CHINA AND THE UNITED STATES[8]

C.U.P.M.		Chinese	
Analysis	3 courses	Analysis	2 six-semester-hourcourses
Abstract			
Algebra	2 courses	"Higher"	
		Algebra	1 four-semester-hour course
Geometry	2 courses	Projective	
(beyond		Geometry	1 two-semester-hour course
Analytics)			
Probability	2 courses	Differential	1 two-semester-
and Statis-		Geometry	hour course
tics		Probability	
Upper Level	2 courses	and Statis-	1 four-semester-
Electives		tics	hour course
		Upper Level	
		Required	
		Courses:	
		Differential	1 four-semester-
		Equations	hour course
		Functions of	1 three-semester-
		Real Vari-	hour course
		able	
		Functions of	1 four-semester-
		Complex	hour course
		Variable	
		Electives:	
		Theory of	
		Functions	
		Electives:	
			Courses to
		Modern Algebra	comprise 200
		Differential	hours of work
		Equations	in fifth
		Applied Statistics	year
		Numerical Computa-	
		tion	
		Applied Analysis	

[8]Ibid., p. 1102.

The comparison reveals that the Chinese curriculum for mathematics teacher-training is equivalent at the elementary level, but exceeds the American standards in the advanced courses required. Both curricula specify approximately twenty-four semester hours of work in mathematics at the elementary level, with the Chinese requirements exceeding the American standards in analysis, and the American exceeding the Chinese in algebra. However, in upper level mathematics work, the Shanghai school requires eleven hours plus work in electives, whereas the C.U.P.M. standards ask for only six hours of electives. While this example of Chinese teacher-training in mathematics appears to exceed C.U.P.M. standards, it must be remembered that the Shanghai University was a prestige institution at the time this program was in effect. Its staff and curriculum were superior by Chinese standards. Thus, comparisons of American programs with the program of the Shanghai school should not be generalized to encompass teacher education programs in the People's Republic as a whole.

Formal training of primary school teachers is given by senior normal schools. All students in a primary program are trained as generalists and undertake a course of study similar to that received by senior middle school pupils.[9]

Informal Teacher Training

Often forced by the demands of crash educational
programs into employing teachers without adequate professional training, the Chinese government has attempted
to provide inservice training for these people in a
variety of ways. Spare-time educational programs and
correspondence studies are made available to most teachers. One of the more interesting of these teacher-
training programs is conducted by television. Television
universities first came into existence in 1960 in the
cities of Shanghai, Peking, Canton, Harbin, and Shen-
gang.[10] Early morning broadcasts originating from local
universities and normal schools were transmitted via
closed circuit television to study centers scattered
strategically around these cities.[11] Viewers received
instruction in four disciplines--physics, chemistry,
mathematics, and Chinese--and were expected to devote

---

[9]For a primary teacher training syllabus, see
Theodore Chen, Teacher Training in Communist China (U.S.
Department of Health, Education, and Welfare, Washing-
ton, D. C., December, 1960), p. 18.

[10]K. E. Priestly, Education in China (Hong Kong:
Dragonfly Books, 1961), p. 32.

[11]Fan Chih-Lung, "College Courses by Television,"
China Reconstructs (April, 1961), pp. 10-11.

eight hours per week to their study, four to viewing
broadcasts, and four to homework. If possible, leading
scholars and well-known teachers present the lectures.
Professor Su Pu-ch'ing of Fu-tan University has been
known to give mathematics presentations.[12] Films and
models are used whenever possible. Selected mathematical exercises stressing production applications are
solicited from local industries and offered as demonstration examples. Readings and homework are assigned at
the end of each lecture. Teaching guidance stations
located about the cities provide student tutoring
services and conduct required testing. The laboratories
of local schools are available for experimental work.
By 1961, it was estimated that 4,100 teachers were par-
ticipating in televised mathematics studies in Peking
alone.[13] Other cities, such as Wushi, have begun to re-
televise the initial lecture series, thus widening the
viewing audience.[14] Four-and-a-half to five years of
participation are necessary for graduation from the

---

[12]China News Service Bulletin, "First Batch of
Students Graduated from Shanghai Television University,"
(Canton, February 5, 1966), p. 5; translated in S.C.M.P.
36313.

[13]Kuang-ming Jih-pao (Peking, November 11, 1961);
S.C.M.P. 2654.

[14]Chih-jih Hsin-wên (Peking, May 7, 1966), p. 7;
S.C.M.P. 36409.

program. Many Chinese teachers can already boast of
having graduated from this unusual program.

The Effects of the Cultural Revolution
on Teacher Training

Since the advent of the Cultural Revolution, the
content of regular teacher-training programs has been
altered by increased emphasis on student political,
military, and production involvement at the expense of
the usual professional course work. The full extent of
these changes is not obvious, but one notable result has
been an increased proletarianization of teacher trainees.
Normal school students are assisting in the thrust to
expand the number of general educational facilities for
workers and peasants by establishing schools in factories
and on communes. They have also been instrumental in
the Mao-inspired revision of text materials. These
experiences are providing China's future teachers with
a valuable practical approach to teaching that may well
compensate for their reduction in formal academic preparation. Since 1964, the emphasis on spare-time normal
training has increased appreciably. Hopeh province
produced a "Draft Education Plan for Four-Year Part-time
Normal Schools," which followed the curricula of the
regular middle school. Secondary mathematics teachers
trained under the program complete 412 hours of mathematical studies of which 192 are devoted to algebra, 74 to

solid geometry, and 128 to arithmetic including 40 hours
spent in agricultural bookkeeping, 15 hours on statistics,
and 18 hours developing computational skills on the
abacus.[15] In 1965, the Mathematics Faculty Research
Section of Ch'ang-show Spare-time Normal School in
Szechwan province published their suggestions for abbreviated normal school studies.[16] They condensed the
former 391-hour, four-year study of principles and concepts of elementary mathematics into a two-year, 284-
hour period. Such innovations, while fashionable during
this period of educational flux, cast some doubt on the
quality of training received by students in spare-time
normal programs.

The Teacher Shortage

There is no question that during the past twenty
years the Communists have responded to their national
need for mathematics teachers in many ways. Attempts
aimed at satisfying the state's requirements for large
numbers of educational personnel have resulted in

---

[15]Yang-ts'un Normal School, Hopeh, "Prelimary
Views on Mathematics Pedagogy in Four-Year Part-Time
Normal Schools," Shuxue Tongbao (August, 1965), pp. 2-6
and 20.

[16]Mathematics Faculty Research Section, Ch'ang-
shou Spart-time Normal School, Chungking, "Arrangement
of Text Material on Mathematics in Spare-time Normal
Schools," Shuxue Tongbao (July, 1965), pp. 2-3.

developing a body of mathematics teachers whose background varies greatly from that of the Shanghai Pedagogical University's trained theorist to the "barefoot" peasant relating his stories of past exploitations. This fact, together with the existing lack of standardized examinations or levels of achievement, is producing a generation of educated Chinese that, although united in its general academic accomplishments, is widely diversified in its mathematical knowledge. On the industrial market this nonuniformity may provide dire consequences. Production expectations cannot be made as flexible as the mathematical background of workers. While the quantity of mathematics teachers has increased, it has not kept pace with the overall expansion of the school system. There is still a critical lack of trained mathematics teachers in the People's Republic of China. In the "Ten Great Years" of the nation's development from 1949 to 1959, normal school output increased by 300 per cent; however, during this same period primary school enrollments increased by 363 per cent, and middle school by 570 per cent.[17] Similar gaps in the supply and demand for mathematics teachers continue to plague the Chinese, and it is most probable that they will continue into the foreseeable future.

---

[17]Chang Chien, "Schooling for the Millions," China Reconstructs (October, 1959), p. 56.

Psychological Research into the Processes of Mathematics Learning

Reconstruction and Reorientation of Psychological Research Facilities

Upon its ascension to power, the Communist government resurrected psychology as a major science by including an office for it in the Chinese Academy of Sciences in 1950. In 1951, the Academy formed the Psychology Research Office, which in five years evolved into the Institute of Psychology. American-inspired theories in psychology were reformulated to conform to Soviet models. This effort incorporated the Marxist-Leninist philosophies of dialectical materialism with the scientific theories of Pavlovianism. For the Soviets, the state of human consciousness is founded upon a socially conceived objective reality, and language as a product of human society becomes inseparable from consciousness. Consciousness will produce "correct" behavior. In principle, a consciousness founded on a socialist exemplar should reflect socialist ideals in its contribution to the state. A cycle is established that parallels Mao's theory of knowledge and is in harmony with the Chinese Communist milieu. Pavlov, while known mostly in the West for his famous work on conditioned reflexes, made his major contributions to the sciences of human activity by investigations

on verbally stimulated human reflexes. This work encompassed the thought processes of abstraction and general-
ization. Pavlov termed these activities of the higher
nervous system the "second signal system." It is this
aspect of Pavlov's work that lends itself to socialist
ideology, both Soviet and Chinese.

The initial activities of Chinese psychologists
and psychological workers involved reorienting themselves
through "thought remolding" and obtaining a background in
Soviet theories and research techniques. By 1956, a
twelve-year scientific development plan had been devised
by the state's leading psychologists. The plan contained
provisions for educational research. In 1957, the
Ministry of Education founded its own Educational Science
Research Institute with a section devoted to educational
psychology. Soon major normal universities formed
departments of educational psychology. By 1960, the
published results of various psychologically oriented
educational experiments and studies began appearing in
Chinese journals.[18] Much of this work was prompted by
the Great Leap Forward's adjurations for scholars to
become involved in the problems of the working classes.
Most numerous among these studies were investigations

_____

[18]Hsin-li Hsüeh-pao [Acta Psychological Sinica].
Chinese Academy of Science, Peking; Shuxue Tongbao.

concerning language and mathematics learning. While this
choice coincided with the state's educational priorities,
it was not necessarily contingent upon them, but rather
rested with the transmitted Soviet research emphasis.
Between the years 1957 and 1960, thirty-three Chinese
studies were done on mathematics learning.[19] These
investigations mostly concerned the "formation of number
concepts, the development of logical thinking and problem
solving."[20] Although Soviet influence was predominant
in this work, theories of some Western psychologists,
particularly Jean Piaget, were known to China's researchers and exerted some influence on their experiments.

Results of Chinese experiments have been more
qualitative rather than quantitative. Researchers seem to
prefer working with small sample sizes, and attempt to
control several experimental variables simultaneously.
Detailed statistical analysis of data is avoided. Some
completed experiments have come under criticism for
arriving at conclusions through the use of inductive
generalizations instead of a systematized approach. At
the December 1963 meeting of the Psychological Society

[19]Lu Chung-heng, Hao Yu-yen, Ying Yu-yeh, Ma
Chieh-wei, and Chang Mai-ling, "Some Psychological Fac-
tors in Promoting the Student's Grasp of Arithmetical
Knowledge, as Found in Recent Educational Reforms,"
Hsin-li Hsüeh-pao, No. 3 (1961), pp. 190-201.

[20]Kuang-ming Jih-pao (March 9, 1962).

in Peking, it was pointed out that research on thinking should proceed from child development, language and logic, to the use of quantitative methods, and recent scientific techniques. The use of a stimulus-response approach allowing one to "grasp two ends and aim towards the middle" was urged.[21]

Research on Mathematics Learning

Of the experiments conducted in the education field, many were concerned with the improvement of classroom teaching efficiency. In the spring of 1960, a team composed of Communist Party representatives, educators, and members of the Institute of Psychology devised an experimental second-year primary curriculum designed to enable pupils to "acquire knowledge systematically and make them think."[22] The materials were introduced at the Beiguan Primary School in Peking. The experimental second-grade arithmetic class successfully coped with numbers up to 100,000, exceeding the usual level of 100, and performed multiplication with two digit numerals, a feat normally expected of fourth grade students. Qualitative results of tests showed the marked superiority of

---

[21]Ch'en Ta-jou, "Annual Academic Meeting of the Chinese Psychological Society," Hsin-li Hsüeh-pao, No. 1 (1964), pp. 109-112.

[22]S.C.M.P. 2270, 1960.

of the experimental group's performance as compared with
the control group:

	Experimental	Control
Mean Score	96.6%	90.2%
Pupils Achieving 100%	54.4%	38.1% [23]
Pupils failing test	0 %	Unknown

Most experiments at this time were conducted with the
government's concern for accelerating the educational
process in mind. Researchers attempted to study how various disciplines could be moved downward in the school curriculum. This practice was particularly true of algebra
studies. A team composed of Huo Mao-cheng, Chang Tung-
chun, and Liu Ching-ho tested the hypothesis that primary
school students would be more proficient in problem solving if given an advanced background in algebra.[24] The
experiment was held at the Peking Second Experimental
School where Huo and Chang were teachers. Liu was from
the Institute of Psychology. Four fifth grade test
groups--A, B, C, and D--were chosen on the basis of a
general examination (see Table 21). Groups A and B and C
and D respectively were of the same ability, with groups
C and D superior to A and B. A and B comprised the experimental samples, C and D the controls. Materials were

---

[23]Ibid; also given in Price, Education in Communist China, p. 116.

[24]Huo Mao-cheng, Chang T'ung-ch'un, and Liu Ching-
ho, "Teaching Algebra in Primary Schools Experimentally,"
Shuxue Tongbao (April, 1960), pp. 14-18.

written that contained twenty hours work in second-year middle school algebra, fifth through seventh-grade word and computational problems usually solved by arithmetic techniques.

TABLE 21

TEST RESULTS OF ALGEBRA SUPPLEMENTED
INSTRUCTION IN FIFTH GRADE PRIMARY
SCHOOL[25]

(Numbers indicate percentage of students achieving specified grades.)

First Test Period

September-October 1958

Grade	1st Quiz		2nd Quiz		3rd Quiz		Gen. Exam.	
	A	B[2]	A	B	A	B	A	B
5	27		59	37	57	61	82	65
4	59		22	39	32	37	16	20
3	14		14	19	7	7	2	15
2			5	5	4	5		
1								

Second Test Period

October-December 1958

Grade	1st Quiz		2nd Quiz		3rd Quiz		Gen. Exam.[1]	
	A	B	A	B	A	B	A	B
5	50	48	74	38	83	65		
4	41	36	17	28	10	30		
3	7	13	5	17	7	5	96	92
2		3	4	17				
1	2							

---

[25]Ibid., p. 16.

TABLE 21 (Continued)

Third Test Period

May–June 1959

Grade	Test 1 5/6/59		Test 2 6/7/59			
	A	B[3]	A	B	C	D
5	88	90	47	42	17	8
4			28	11	11	22
3			11	3	22	26
2			6	3	6	22
1			8	41	44	22

1. Represents the groups average grades.
2. No specific number given--all students were said to pass.
3. Represents the groups average grades.

All test groups studied the same materials; however, their methods of instruction differed. Group A learned using a concrete approach. Group B's instruction stressed the learning of algebraic formulae. Groups C and D were taught by the usual lecture method. Three periods of testing were employed, two of which only concerned groups A and B and tested their relative progress in mastering the materials. The third test period involving all four groups was given after a duration of four months from exposure to the experimental materials and was intended to measure respective retention of knowledge and techniques learned. The researchers use a Soviet five-point grading system where a five indicates superior performance and a three average. Clearly, group A taught through a concrete approach achieved the best performance.

The researchers concluded that middle school algebra could successfully be taught at the primary level and would facilitate solving complex arithmetic problems. Work of a similar nature was done by Ch'en Ch'uan-hui, Sun Ching-hou, Yang Hung-ch'ang, and Chou Husi-shu at the same school.[26] Ch'en was on the staff of the Peking Second Experimental School while his colleagues were members of the Institute of Psychology. Using small samples, nine students, their experiments demonstrated it was possible to teach material involving second-degree algebraic equations in the fifth year of primary school, and to teach third year students how to find the roots of given numbers. Both topics had previously been reserved for middle school instruction. Still another team composed of Institute psychologists and a school teacher proved that calculus could be taught to fourteen- and fifteen-year-olds in the junior middle school.[27] The results of these experiments were instrumental in the

[26]Ch'en Ch'uan-kui, Sun Ching-hao, Yang Hung-ch'ang, and Chou Huai-shui, "Experimental Teaching and Results of 'Second Degree Equations' of Algebra in Fifth Year Primary School Class," Shuxue Tongbao (June, 1960), pp. 7-11; "The Experimental Teaching and Results of 'Radical Numbers' of Algebra in Third Year Primary Class," Shuxue Tongbao (June, 1960), pp. 5-7.

[27]Yang Yu-men, Ko Chien-jen, Chou Huai-shui, Sun Ching-hao, and Yang Hung-ch'ang, "Experimental Teaching and Results of 'Limits' of Algebra in Second Year Junior Middle School Class," Shuxue Tongbao (June, 1960), pp. 11-13.

proposals for mathematics teaching reforms suggested by the Peking Normal University in 1960, and were specifically referred to in speeches at the National People's Congress in connection with educational reforms.

Investigations on geometry teaching were undertaken by psychologists such as Lu Chung-heng. Lu's work included "An Experimental Study of the Effects of Different Methods of Teaching on the Formation of Fundamental Geometrical Concepts,"[28] "The Negative Effects of Crossing of Geometrical Figures on the Perceptual and Thought Processes,"[29] and "The Effects of Outer Shape on the Perceptual Structure of Geometric Figures and Thought Processes."[30] This last experiment, involving children from an experimental school in Liaoning, centered on a Pavlovian "second signal system investigation." In his work on mathematics learning, Lu has been heavily influenced by the writing of the Soviet researcher Rubenstein.[31] Studies concerning thought processes and

---

[28]Robert Chin and Ai-li S. Chin, Psychological Research in Communist China: 1949-1966 (Cambridge, Mass.: M.I.T. Press, 1969), p. 145.

[29]Ying Yu-yeh, Chang Mei-ling, Chu Hsin-men, Hung Tie-lun, and Yen Hwei-fen, "The Negative Effects of Crossing of Geometrical Figures on the Perceptual and Thought Processes," Hsin-li hsueh-pao, No. 1 (1964), pp. 72-93

[30]Lu Chung-heng and Chu Hsin-ming, "The Effect of Outer Shape on the Perceptive Structure of Geometric Figures and Thought Processes," Hsin-li hsueh-pao, No. 3 (1964), pp. 248-257.

[31]Chin and Chin, op. cit., p. 145.

arithmetic operations were performed by Ch'en Li of the
Hangchow Pedagogical Institute. One such study concerned
the process of abstraction and the development of an
"active nature of thought" in the understanding of the
concepts of inequality.[32] Another tested the feasibility
of teaching an operation--the inverse approach as opposed
to merely teaching the operation. It was found that this
method was indeed superior for use in problem solving,
and increased the student's "flexibility of thinking."[33]

Several of the previously mentioned experiments
were undertaken after 1960. This post-1960 era, in
particular the years between 1964-66, marks an extremely
productive period for mathematical-psychological research
in the People's Republic. By now the merits of research
in educational psychology were appreciated, and a body of
trained researchers organized who built upon the experience gained from the research efforts of the Great Leap.
This body of researchers now became involved in the
national effort to improve the quality of education.
Early in 1961, the Chinese Psychological Association and

[32]Lu Ching, and Wang Wen-chun, "The Thinking
Processes in Arithmetical Operation in Lower-Grade
School Children," Hsin-li Hsüeh-pao, No. 2 (1960), pp.
121-135.

[33]Lu Ching, Wang Wen-chun, and Cheng Yu-tan, "The
Mastery of Reverse Operation and Flexibility of Thinking
in Arithmetic in School Children," Hsin-li Hsüeh-pao,
No. 3 (1964), pp. 237-247.

several other research-oriented organizations held a
series of lectures in Peking for school teachers.[34]
This series was intended to introduce the teachers to
the problems, methods, and functions of educational
research and, if possible, get them involved in research
efforts. In a speech at the Educational Psychology Conference in 1962, P'an Shu, spokesman on experimental
work, indicated the directions investigations were taking. Included in his list of problems under consideration were:

1. Psychological aspects of classroom instruction that would improve the learning process
   in relation to motivation, acquisition of
   skills and the development of logical thinking in students.
2. Improved coordination between the learning
   activities of informal preschool education
   and formal school education.
3. The determination of and allowance for individual differences in student learning and
   the development of corrective procedures for
   learning deficiencies.[35]

Researchers responded to these tasks by undertaking a
variety of experimental projects. In mathematics learning some experiments were of a very general nature,
others concentrated on specific problems.

---

[34]Kuang-ming Jih-pao (Peking, March 8, 1961).

[35]Pan Shu, "Ideas about Expanded Research in
Educational Psychology," Kuang-ming Jih-pao (March 13,
1962); Joint Publication Research Service No. 13531.

One Piagetian-type experiment involved an investigation into the conceptualization ability of children four to nine years old.[36] The three investigators were concerned with isolating the developmental trends reflected in children's ability to classify, and to determine the ages at which classification could be viewed as a turning point towards conceptual thinking. Their findings showed:

1. Children under four years old fail to carry out classification.
2. Children in the five-six category demonstrated a rudimentary ability to classify based mainly on perceptual activity.
3. Children six-seven showed some ability in conceptualization.
4. The turning point from perceptual to conceptual thinking appears to fall between the ages of five and six.

A similar experiment also involving fundamental learning investigated the development of primary school children's ability to compare.[37] It was found that this ability was contingent upon age level, the nature of the objects used, and the methods employed for comparison.

Perhaps the most recurrent type of investigation concerned children's difficulties in obtaining solutions

[36]Wang Hsien-tien, Liu Cheng-ho, and Fan Tsun-jen, "An Investigation into the Development of Concepts in Children 4-9 Years [Old]," Hsin-li Hsüeh-pao, No. 4 (1964), pp. 352-360.

[37]Wei Chang, Hwang Hsio-ying, Sung Ching-yao, and Ch'i-fen, "The Development of the Ability to Compare in Primary School Children," Hsin-li Hsüeh-pao, No. 3 (1964), pp. 274-280.

to verbal mathematics problems. Hsiao Chien-ying
attempted to isolate the characteristics of the thought
processes of first-grade children involved in solving
verbal arithmetic problems. He found that their thinking moved from a concrete imagery towards abstraction
and that seven-year-olds were capable of limited abstract
thinking.[38] Cheng Tsu-hsin, working with second grade
students, conducted a study to determine the effects of
children's activity on verbal problem solving.[39] He
found that external activity helped lower-grade children
in understanding and solving verbal problems. In still
another study entitled "The Relation Between the Structure and Formulation of the Arithmetical Problems and
the Psychological Activity of the Pupils in the Process
of Problem Solving,"[40] the relation between maturation
and problem-solving ability was further affirmed. In
refining these theories, Chen Pei-lin and Mao Yu-yen

---

[38]Hsiao Chien-ying, "The Characteristics of the
Thinking Process of Solving Verbal Arithmetical Problems
by First Grade Children," Hsin-li Hsüeh-pao, No. 1
(1965), pp. 50-56.

[39]Cheng Tsu-hsin, "An Experimental Study of the
Effect of Children's Activity on Solving Verbal Problems
in Division," Hsin-lin Hsüeh-pao, No. 4 (1964), pp. 369-
376.

[40]Mao Yu-yen, Kung Wei-yao, and Chen Pei-lin,
"The Relation Between the Structure, Formulation of the
Arithmetical Problems and the Psychological Activity of
the Pupils in the Process of Problem Solving," Hsin-li
hsüeh-pao, No. 4 (1965), pp. 291-297.

found, in sampling the problem-solving ability of fifth-
grade pupils, that there were positive correlations
between the students' achievement in classifying, solving,
and composing problems. They also found that six levels
of understanding and mastery in problem solving could be
isolated.[41] A full application of the results of such
experiments in curricular and pedagogic improvement was
prevented by the advent of the Great Cultural Revolution.
What effects the experimental conclusions will have on
post-Cultural Revolution teaching remains to be seen.

A few research projects reflect problems that are
distinctly Chinese. Ch'iu Hsueh-hua of the East China
Normal University isolated three stages of physiological-
psychological mathematics learning on the abacus. These
states were (1) use of audible verbal signals, (2) silent
verbal calculations, and (3) automatic manipulation.[42]
Experimenting within a Socialist framework, Liu Ching-ho
has identified a "superior conception of number" in the
children of peddlers, as compared with professors' children and those of office workers.[43] Liu attributed this

---

[41]Chen Pei-lin and Mao Yu-yen, "On the Processes of
Mastery of Typical Verbal Arithmetical Problems in School
Children," Hsin-li Hsüeh-pao (No. 3, 1965), pp. 215-222.

[42]Ch'iu Hsueh-hua, "A Study of the Skill of Calculation with Abacus in School Children," Hsin-lin Hsüeh-
pao (No. 2, 1963), pp. 107-112.

[43]Liu Ching-ho, "A Preliminary Discussion on Some
of the Problems Regarding the Learning Process of Children," Jen-min Jih-pao (March 26, 1961); J.P.R.S. No. 9398.

phenomenon to the "past experience" of the peddler's
children with quantities.

All of this experimental work indicates that
Chinese educators are well aware of the necessity for
psychological research into the processes of mathematics
learning and teaching. The existence of experimental
schools implies a commitment to research, and the geographical distribution of institutions performing
investigations in this area gives evidence of the nation-
wide concern for teaching improvement prior to the Great
Cultural Revolution. In light of the revolutionary
changes that have taken place in Chinese education since
1966, it is most likely that such research has been de-
emphasized and classified in the broad category of
bourgeois activities.

Extracurricular Mathematics Activities

A vital aspect of Chinese education from the mid-
fifties onward has been the utilization of extracurricular
activities in the promotion of priority subjects. This
concept was adopted from the Soviets and used with great
advantage in the People's Republic. Up until the time
of the Cultural Revolution, the discipline to receive the
most extracurricular attention was perhaps mathematics.
Originally the government suggested that schools and
teachers set up such a program, but gave no specific

outlines as to how it was to be set up. Different
schools satisfied their students' needs in varied ways.
The Young Pioneers, of which there was a troop in every
school, formed many clubs, including ones devoted to the
study of mathematics. After the establishment of the
national mathematics competitions, competition study
groups were formed among students in many large city
schools. Some teachers organized collective study groups,
one for each mathematics subject: elementary algebra,
plane geometry, trigonometry, and advanced algebra.[44]
These groups, really comprising after-school "help sessions," were led by brighter students and were intended
to serve as a source of assistance for their less proficient colleagues.

Student Mathematics Study Groups

The activities of study groups, or student research
teams, as they were sometimes called, were designed to
stimulate the participants' interest in mathematics and,
in general, improve their schools' overall mathematics
standards. The Second Middle School of Peking Normal
University involved its seventh- and eighth-grade students
in activities related to stories concerning mathematics,
such as the discovery of Neptune; tenth-graders received

---

[44] Huang Sung-men, "On After-hour Mathematics Study Groups," Shuxue Tongbao (November, 1964), pp. 17-22.

after-school lectures by teachers on the theory of equations; the eleventh grade engaged in bi-weekly projects
related to their immediate classwork.[45] Ninth- and
twelfth-grade students were excused from after-school
activities so that they could prepare for their final
examinations (see Table 22). Some study groups edited
and published school-wide mathematics newspapers to the
benefit of all their classmates.[46] Others specialized in
a variety of mathematics-related activities, including
abacus popularization, chart making, accounting study, the
construction of teaching aids, and survey practice.[47]
This latter approach combined academic studies with
practical applications.

---

[45]Mathematics Faculty Research Section, Second
Middle School, Peking Normal University, "Conducting
Active Extracurricular Activities," Shuxue Tongbao
(February, 1963), pp. 2-4.

[46]Wang Chih-heng, "Our Mathematical Land," Shuxue
Tongbao (September, 1964).

[47]Wu-Gong, "To Realize Educational Policy from
Extracurricular Activities," Z'hongxue Shuxue (September, 1959), pp. 7-8.

TABLE 22

SCHEDULE OF ACTIVITIES FOR MIDDLE SCHOOL
MATHEMATICS STUDY GROUPS[48]

Week of Term	Ninth Grade	Eighth Grade	Seventh Grade
4	Organization of work Choose leaders for sections	Same	Same
5	Study problems related to algebra	Discuss characteristics of triangles	Work interesting problems involving fractions
6	Manufacture surveying instruments; edit a blackboard report	Construct teaching aids and charts	Construct statistical charts
7	Study selected geometrical problems	Manufacture models concerned with square roots	Same
8	Do the five circle tangent problem; edit a blackboard report	Construct models concerning cube roots	Construct models
9	Mathematical games	Same	Same
10-11	Field work; blackboard reports	Same	Same
12	Outside survey work; height of a tree, width of a river	Same	Indoor and outdoor survey concern-area and volume
13	Outside survey; cartography	Construct models of cylinders and cones; edit blackboard report	Same

[48]Hung Guang-shun, "Extracurricular Mathematics,"
Shuxue Tongbao (May, 1956), pp. 39-40.

TABLE 22 (Continued)

Week of Term	Ninth Grade	Eighth Grade	Seventh Grade
14	Construct charts illustrating polygons	Study construction of triangles	Free activity
15	Discuss native and foreign mathematics and applications of mathematics	Same	Same
16	Study short cuts for mental, written and oral calculation	Edit blackboard report	Study calculation short cuts
17	Discuss selected problems in geometry	Compare solution methods for algebraic problems	Discuss applications of proportion
18	Summarize semester's work	Same	Same

Mathematics Parties

A very unusual after-school activity was the mathematics evening party. An article in the October, 1962, issue of Shuxue Tongbao, by V. G. Kolatsunowa, described such a party at the Moscow Middle School No. 710.[49] Kolatsunowa's party consisted of the presentation of an oral report on the biography of a famous Russian mathematician and a mathematical contest among those present,

---

[49]V. G. Kolatsunowa, "The Procession of a Mathematics Evening Party," Shuxue Tongbao (October, 1962), pp. 7-8.

with prizes being awarded the winners. The Chinese version suggested by Wu Ch'i-ch'i of T'ai-shan Middle School
No. 1 would seem more appealing to a general audience.[50]
His activities included a chorus of mathematical songs,
recitation of mathematical poems, a performance of comic
monologues on a mathematical topic, mathematical games,
lectures on the history of mathematics, and a mathematical
play. A typical game played at this party would involve
two teams of fifteen members each and an umpire.[51] Individual team members would represent the numbers 1 to 10
and the symbols -, +, -, and ÷ and x. On the umpire's
command, giving an arithmetic expression, i.e. (2 x 4) -
8 = 0, each team would attempt to order its members to
correctly represent the designated mathematical statement.
The first team to accomplish the feat won the round.

Benefits of Extracurricular Work

There is little doubt that the additional student-
time spent in extracurricular mathematical activities
assists in improving the academic performance of Chinese
students. In their proposing of the scheme, educational

[50]Wu Ch'i-ch'i, "The Contents of Mathematics Evening Parties," Shuxue Tongbao (October, 1962), pp. 6-7.

[51]Edward Hunter, Brainwashing in Red China (New
York: Vanguard Press, 1951), p. 46.

officials hoped that the success of after-school mathematics sessions would induce a percolation effect that
would assist in raising the standards of mathematics
instruction.[52] Theoretically, a chain-reaction was begun
with enthusiastic study group members stimulating their
fellow students to undertake more rigorous mathematical
studies. Thus, with student interest elevated, teaching
instruction would be forced to improve in quality and
content. While broadening and strengthening the students' mathematical knowledge, extracurricular activities
also supply a rejuvenating freedom from the regimentation
of classroom study and cultivate an appreciation for
independent study. Chinese students have an affinity for
collective endeavors and work most efficiently in such
group situations.[53] Although designed mainly to promote
the state's educational goals, extracurricular activities
serve a practical function for China's burdened teachers.
They assist teachers by

1. Providing remedial work and assistance for
   slower learners.

2. Supplying a ready source of teaching aids--
   charts and mathematical models that could
   be utilized for the instruction of all
   classes.

---

[52]Fu Tsu-ch'ung, "Middle School Extracurricular
Mathematics Activities," Shuxue Tongbao (April, 1963),
pp. 2-4.

[53]Observed by teaching Chinese students in Malaya,
1964-67.

3. Serving as a means of supplementing theory
    with "practice."

In the wake of the Great Cultural Revolution, the
emphasis of such activities has been directed towards
furthering student political development. Middle school
after-hour exercises now stress study of Chairman Mao's
works, military sports practice, and the acquisition of
proletarian skills, such as barbering and sewing.[54]
Students are urged to serve and inspire the working
classes by singing revolutionary songs, presenting
patriotic skits, and propagating Mao's thought in factories and villages.[55] These "Red Little Soldiers" are
often led and instructed by local cadre, old workers,
or former army personnel who live in their neighbor-
hood.[56] Mathematical activities are still pursued, but
on a minor scale as compared to the previous pre-1967
period.

---

[54]Aomen Road No. 2 Primary School, "Shanghai
Puts Extracurricular Activities on the Agenda of Its
Revolutionary Committee," Jen-min Jih-pao (July 10,
1970); S.C.M.P. 4706.

[55]"Using Mao Tse-tung's Thought to Occupy the
Positions Outside the School," Surveys of China Mainland
Press (June 10, 1970), No. 4673; "Students Assisted in
Their Creative Study and Creative Application of Chair-
man Mao's Works Outside the School," Kuang-ming Jih-pao
(January 14, 1969), pp. 3-4; S.C.M.P. 4351.

[56]"Grasp Extramural Education," Jen-min Jih-pao
(August 27, 1969), pp. 1-3; S.C.M.P. 4492.

The Chinese Mathematical Olympiads

Two articles in the January 1956 issue of Shuxue
Tongbao advocated the establishment of national mathematical examinations in the People's Republic of China
patterned after the Soviet Mathematical Olympiads. The
first article by Professor Hua Lo-kêng, Director of the
Institute of Mathematics of the Chinese Academy of
Science, expressed his commitment to the establishment
of such a scheme in China.[57] During a visit to Moscow
in April of 1946, Professor Hua was impressed by the
enthusiastic response given by middle school students to
a lecture on complex numbers by P. S. Aleksandrov of
Moscow University. These students were members of study
groups preparing themselves for participation in the
Soviet Mathematical Olympiads. Hua visited Russia again
in 1953 as a member of a delegation from Academica
Sinica, and was advised by Soviet educators to initiate
mathematical competitions in China as a method of promoting scientific advancement. It was felt that through
such activities Chinese youth would be stimulated toward
mathematics studies, thus forcing an improvement in the
quality of school mathematics and science instruction.

---

[57]Hua Lo-kêng, "We Will Have National Mathematics
Competitions Soon," Shuxue Tongbao, Chinese Mathematical
Association, Peking (January, 1956), pp. 1-3.

Firmly convinced of the potential national benefits of
mathematical competitions, Hua suggested their adoption
as an extracurricular activity, but also cautioned
against a resulting disruption in the regular school
system. The examinations were not to interfere with the
school's normal functions. In the second journal article,
Tuan Hsueh-fu, Professor at Peking University, urged
Chinese educators to "learn from Russia" concerning
mathematical competitions.[58]

Organization and Execution of the Examination Scheme

Activities soon began in earnest to implement Hua
and Tuan's recommendations. Mathematics competition
committees were established in Peking, Shanghai, Tientsin,
and Wuhan. They were responsible for local organization
of contests and for setting examination questions.
Shanghai's committee was composed of seventeen members
selected from the Mathematical Society, the Shanghai
Municipality Education Office, and the local chapter of
the Chinese National Association of Natural and Special
Sciences. In the choosing of examination questions, the
committees were required to select topics from arithmetic, algebra, geometry, and trigonometry that, while

---

[58]Tuan Hsueh-fu, "Learn from Russia to Have Mathematical Competitions," Shuxue Tongbao (January, 1956), pp.
3-5.

248

difficult, did not exceed the level of work required by middle school mathematics outlines. Similar to the Soviet scheme, associated student lectures on some aspect of mathematics were to be given. The first Peking lecture was given in the afternoon of March 11 by Professor Su Pu-Chin. His topic was "Non-Euclidean Geometries." The Tientsin lecture for March was given by Hua Lo-keng.[59]

In May the first examination was undertaken. Students in the last two years of middle school were given a screening examination by their teachers. Those who did well and were politically acceptable were recommended to represent their schools in city-wide competition. The official examination was composed of two rounds, with the final winners emerging from the second round. Each round contained five or six problems in a given time allotment of 150 minutes. Students who passed the first round were awarded a certificate of merit and allowed to compete in the second round. Success at the last level warranted a medal and an award of books. The competitors with the three best scores were permitted entrance to the universities of their choice to study either mathematics, physics, astronomy, or any other associated scientific discipline, without being subjected to

_____

[59]Shuxue Tongbao (April, 1956), p. 2.

further examinations. Naturally, the accomplishment of
doing well in such an examination brought great recognition to the young scholar, and for a short period he
became a local hero much like the successful civil service candidates of old. On May 4, Wuhan conducted its
examinations and had twenty-one students pass. Sixty-two
Peking middle schools sponsored 622 students in the final
round of its competition of May 13. Thirty-three passed.
Tientsin's examination on May 27 had 499 participants in
the final round with twenty-five passes. Shanghai's
Olympiad was given in early June and saw 732 contestants
in the second round (no information is available as to
the number of final winners).[60] Although the examination
efforts in these four cities were considered experimental,
they were acclaimed outstanding successes. Shanghai's
experiences of 1956 and the following year, 1957, were
well documented and published to serve as a guide for
other cities to follow.[61]

---

[60]Hua Lo-keng, "Completion of the Peking Competition," Shuxue Tongbao (June, 1956), pp. 1-2.

[61]Shang-hai shih, 1956-57 nien Chung hsüeh-sheng
Shu-hsüeh Ching-sai his-t'i pien-hui [Compilation of Prob-
lems from 1956-57 Mathematics Competitions for Middle-
School Students in Shanghai Municipality] (Shanghai: New
Knowledge Press, 1958). For a fuller discussion of this
work, see John De Francis, "Mathematical Competitions in
China," The American Mathematical Monthly (October, 1960),
67: 756-762.

One hundred thirty thousand copies of <u>Compilation</u> <u>of Problems from the 1956-57 Mathematical Competitions</u> <u>for Middle-School Students in Shanghai Municipality</u> were published and distributed in 1953. In this booklet, the ultimate objectives of the competitions were specified: to locate mathematically talented students so they could be singled out for special educational attention and to encourage self-study and a competitive spirit among students. Both objectives were intended to raise the quality of mathematical training for Chinese students so that the People's Republic of China could compete, scientifically, with the more developed nations of the world. As a result of the examinations, several deficiencies in the student's background were noted and corrective actions suggested:

1. Many students are deficient in analyzing a problem on the basis of hypothesis and conclusions, getting at the heart of the problem and then applying logical reasoning to its solution. Students should be given more opportunity for independent analysis and synthesis. It would be well for them to think about some problems for several days and not to be given the solutions too quickly.

2. Students should be given more practice in handling problems which involved a synthesis of algebra, geometry, and trigonometry.

3. More attention should be devoted in middle school to the study of inequalities so as to facilitate the transition from middle school to college mathematics.

4. Students pay insufficient attention to re-
    stating problems and are inadequately skilled
    in performing operations. The latter was
    especially apparent in complicated problems.
    Hence students should be given comparatively
    complicated problems for homework so as to
    develop their operational skills.[62]

As a result of the competitions, mathematics study
groups were formed in many schools. Students engaged in
extracurricular activities designed to improve their
performance on up-coming examinations. Study groups
existed on several levels: within schools, among several
schools, and at the city-wide level. By 1962, the Peking
Mathematics Study Group boasted a membership of 700.
Members came together once a month to hear a lecture by a
prominent mathematician and to engage in discussions concerning his presentation.[63] Often the lecturer would
pose specific problems to be solved by his audience. In
1960, the Office of Mathematics, Physics and Logic of
the Institute of Mathematics of the Chinese Academy of
Science organized a series of twenty lectures to be
presented in future months and designed for student study
groups. These lectures centered on four themes:

1. An introduction to the study of mathematical
    foundations.

---

[62]Ibid., p. 758

[63]Han Erh-Tsai, "They Like Mathematics," China
Reconstructs, Peking (December, 1962), 11: 34-35.

2. Outline of the history of mathematics.

3. The nature, methods, and significance of
   mathematics.

4. The techniques and characteristics of modern
   mathematics.[64]

Eventually, many of the lectures were published in
pamphlet form for further and more widespread study by
student groups. This Series of Mathematics for Youth
included the following works:

Hua Lo-keng, Discussions Starting from π of
             Tsu Ch'ung
             Discussions Starting from the
             Triangle of Yang Hui

Su Wen-chun, Applications of Mechanics in
             Geometry

Shih Chi-huai, Averages

Tuan Hsueh-fu, Symmetry
               Induction and Deduction

Min Szu-hao, Lattice Points and Area

Chiang K'en-ch'eng, One Stroke Diagrams and the
                    Mailman's Route

Tseng K'en-cheng, One Hundred Mathematical
                  Problems

Ch'ang Keng-che and Wu Jun-sheng, Complex Numbers
                                  and Geometry[65]

Hua's books develop modern mathematical concepts by building upon an historical perspective, thus fostering a
nationalistic and socialist pride in the reader. In the

---

[64]"Office of Mathematics, Physics and Logic of the
Institute of Mathematics of the Chinese Academy of Science
Sponsors Lectures on Mathematics Foundations," Shuxue
Tongbao (February, 1960), p. 42.

[65]Shuxue Tongbao (September, 1962), Back cover.

text on π, he discusses the work of Tsu Ch'ung (429-500),
who at an early date approximated π between 3.1415926
and 3.1415927. Hua develops a series approximation for
π and generalizes series approximation techniques to
analyze periodic phenomena. Orbiting of a Soviet satel-
lite is used as an example. Similarly, Yang Hui's work
with the "Pascal Triangle" (ca. 1261-1275) is promoted
by Hua to a consideration of combinatorics. In his
booklet on geometry, Wu, eminent topologist and academi-
cian, gives the reader a feel for applied mathematics.
Averages by Shih, a teacher at the Peking University of
Science and Technology, progresses from a consideration
of simple everyday problems to the uses of inequalities.
Tuan's Symmetry takes its reader from the perception of
concrete patterns to the abstract theory of permutation
groups. His references of Weyl and Speiser[66] attest to
the level of the material presented. Chiang's book
deals with network analysis and problems in elementary
topology. Tseng's problem book carries a traditional
Chinese flavor. Interestingly written and at a sufficient level of rigor to be demanding, these works served
as an additional source of mathematical stimulation for
Chinese students.

---

[66]H. Weyl, Symmetry (Princeton: Princeton University Press, 1952); A. Speiser, Theorie der Gruppen von
endlicher Ordnung (Basel: Birkhäuser, 1956).

These lectures and publications were part of a
broad government-sponsored campaign to promote the study
of science. At the forefront of this campaign was Hua
Lo-kêng. Mathematician of world renown, concerned
teacher, and confirmed advocate of the Communist Party's
policies, Hua was to be emulated as the socialist model
of a scholar-scientist. The story of his proletarian
background and "Horatio Alger" rise to success despite
adversity was communicated to the youth of China with the
hope that it would encourage them to be persistent in
achieving their educational ideals. The People's Publishers in Shanghai printed his biography, The Mathematician Hua Lo-keng,[67] and Hua, himself, wrote To a Young
Mathematician,[68] in which he included autobiographical
sketches and encouragement to students. Hua was indeed a
self-made man worthy of admiration. Although lacking
higher academic degrees, he has written several classical
works of mathematics and is a versatile researcher
and world-recognized authority in number theory,
harmonic analysis of functions of several complex

---

[67]Shu-hsüeh-chia Hua Lo-keng, The Mathematician
Hua Lo-keng (Shanghai: People's Publishers, 1956).

[68]Hua Lo-kêng, To a Young Mathematician (Shanghai:
China Youth Press, 1956.

variables and group theory.[69]   As a teacher he is deeply

involved in developing the mathematical talent of Chinese

youth.   For Hua, the teacher's function is to act as a

guide for the students.   He advocates a close teacher-

student relationships and condemns learning by rote

memorization.   In many government campaigns, Hua becomes

a spokesman for the scholar scientists.   In particular,

during the early sixties, when the slogan "theory must be

combined with practice" was in vogue, Hua published two

interesting studies: "Applications of Mathematical

Methods to Wheat Harvesting,"[70] which concerned opera-

tions research, and "On the Problem of Calculating

Mineral Reserves and Hill Area On Contour Line Maps."[71]

Results of the Examinations

In subsequent years since 1956, the level of

achievement on the competitions has increased.   This

record is due largely to the influence of student mathe-

matics study groups.   The 1962 competitions in Peking

---

[69]Some of his publications include: _Additive
Prime Number Theory_ (Chinese Academy of Sciences, Peking,
1953); _Harmonic Analysis of Functions of Several Complex
Variables in Classical Domains_ (Izdat. Inostr. Lit.,
Moscow, 1959); _Classical Groups_ (Shanghai Science and
Technology Press, 1963) (with Wang Yuan).

[70]Hua Lo-kêng _et al._, "Application of Mathematical
Methods to Wheat Harvesting," _Acta Mathematica Sinica_,
1961, 1: 77-91.

[71]Hua Lo-kêng and Wang Yuan, "On the Problem of
Calculating Mineral Reserves and Hill Area on Contour
Line Maps," _Acta Mathematica Sinica_, 1961, 1.

attracted 1,465 students from 100 schools, 693 seniors
and 772 juniors representing 5 per cent of their respective grades city-wide. On the first round nearly half
of the seniors scored above 60 per cent correct. The
second round was quite difficult, but one student did
solve all the required problems.[72] Of the eighty-two
eventual winners, half were members of the Peking Mathematics Study Group.[73] From data available on the 1963
competitions, it appears that all student participants
took both examinations, rather than being screened out
by the first round. The examination was later criticized
as being very difficult.[74] Examinations similar to this
one were taking place in Peking up until 1964. No
information is readily available on the Chinese Olympiads
after this period.

It was originally hoped that the mathematical
competition schemes would eventually be adopted by all
large cities in China. Although the movement did spread
from the four cities that inaugurated the tests, it did

---

[72]"Conclusions of the 1962 Mathematics Contest
Among Middle School Students in Peking Municipality,"
Shuxue Tongbao (April, 1963), pp. 50-51.

[73]Han, op. cit., p. 35.

[74]Chao Ts'u-keng, "On the Problems Adopted for the
1963 Mathematics Contest for Peking Middle School Students," Shuxue Tongbao (July, 1963), pp. 8-14; the questions for this examination are given in Appendix D.

not achieve the momentum expected. Perhaps in many
locales, the mathematical talent and organizational
ability for such an endeavor were lacking. The era of
"anti-championism" in the sixties and the Great Cultural
Revolution terminated the examinations. Under pressure
from the red guards, Hua had to publicly confess his sin
of promoting "advanced experience from abroad" in the
People's Republic.[75] The competitions were denounced as
contributing to elitist education practices by encouraging personal achievement. In the years between 1956 and
1964, the existence of the competitions did much to
mould the mathematical thinking patterns of Chinese students. The questions stressed creative thinking over
rigid solution methods dictated by rote-learning experiences. Thousands of students benefited from this exposure. Now in the wake of the Great Cultural Revolution,
it remains to be seen if the educators in the People's
Republic of China will consider this fact important
enough to resurrect the mathematical competitions.

Mathematical Publications

In China, the land where commercial printing
originated, the printed word has long been a source of

---

[75]Hua Lo-kêng, "Chairman Mao Points Out the Road
of Advance for Me," China Reconstructs (November, 1969),
pp. 30-31 and 41.

entertainment and enlightenment. For centuries, small tomes, usually consisting of a few pages of block prints, have been sold by wandering merchants and in stalls in the market places. This tradition has continued into modern times, with the contents of such publications reflecting the Chinese people's thirst for new knowledge.

Now, under the Communist government, this type of book provides a primary avenue of information dissemination. Voluminous quantities of printed materials are produced by government printing houses and stocked in state-owned bookstores. These bookstores serve as a gathering place for China's young, where they spend many hours reading material off the shelves, a practice unappreciated in the West but common in China and other parts of Asia.[76] Supplementing the system of state stores are thousands of privately-owned book stalls dealing in new and used publications. Together these two agencies form a vast avenue for the distribution of printed matter to China's masses.

Much of the resulting traffic in publications deals with science and mathematics materials. Excluding formal textbooks, this literature encompasses teacher's guides and reference works, texts designed for self-study,

---

[76]Photographs of this phenomenon are available in R. F. Simpson's, "The Development of Education in Mainland China," Phi Delta Kappan (December, 1957), p. 89.

professional journals, enrichment-oriented student reading material, and peasant handbooks. Chinese teachers
rely heavily on the counsel of teaching guides and standard reference works in their preparation for classroom
teaching. These guides may concern the intricacies of
specific disciplines or various aspects of teaching,
such as grading examinations or constructing teaching
aids. In mathematics teaching, one of the major Chinese
reference works for many years and still very popular is
the five-volume tract translated from the Japanese of
Nagesawa Kamenosuke.[77] The series includes a volume on
arithmetic, two on geometry, and one each on algebra and
trigonometry. Detailed solutions and explanations for
thousands of mathematical exercises a teacher might
encounter comprise its contents. In the early days of
the People's Republic, Soviet translations provided the
majority of references. These translations included[78]
Estomila's Classroom Teaching Plans for Junior Middle
School Geometry, translated into the Chinese in 1954,
and Classroom Teaching Plans for Junior Middle School
Algebra, 1955; Chichigin's Teaching School Arithmetic,

[77]John De Francis, "The Mathematics Scene in
China," The Mathematics Teacher (April, 1962), 55: 251-
255.

[78]"Reference Books Used in the Past Ten Years,"
Shuxue Tongbao (October, 1959), pp. 26-29.

1954; Serlifev's <u>Teaching Middle School Arithmetic</u>, 1957;
and A. S. Pshyolko's <u>Methods of Teaching Arithmetic in
Elementary School</u>. Starting in 1954, Chinese authors
resumed the task of publishing their own reference works.
The pedagogical techniques advocated by these references
were closely adhered to by readers. In some instances,
they served as texts for normal school students.

Specific government-sponsored educational campaigns
were marked by the appearance of appropriate literature,
often in the form of teacher's reference guides. Thus
the experiences of China's first student mathematical
competition were well documented in <u>Compilation of Problems from the 1956-57 Mathematical Competitions for
Middle School Students in Shanghai Municipality</u>,[79] and
widely distributed to provide information necessary for
the establishment of a national examination system.
Eventually, a special series of booklets designed to
engage student interest in the activities of the competitions were also published.[80] The Party's increased
emphasis on informal education in the mid-sixties was
heralded by the appearance of <u>Analysis of Correspondence</u>

[79]<u>Compilation of Problems from 1956-57 Mathematics
Competitions for Middle-School Students in Shanghai
Municipality</u> (Shanghai: New Knowledge Press, 1958).

[80]<u>Series of Mathematics for Youth</u>; see section on
Mathematical Olympiads for complete list.

Teaching of Mathematics in the Soviet Union for Forty
Years[81] in Chinese bookstores. In such a manner, an
educational policy void, due to the cessation of activities by the Ministry of Education, was prevented.

Self-Study Materials

Perhaps the most popular of all inexpensive books
on mathematics were those intended for self-study. Production of these books varied with the state's campaigns
for part-time and spare-time education, and it would
appear very likely, under the present policy of complete
proletarian education, that they are much in demand.
Contents of self-study books are basically the same as
those found in equivalent school texts, except that the
presentation is less theoretical and many more practical
problems are included for solution. Material is considered
that has immediate relevance for the reader. Self-study
books are usually published as part of a series. One of
the more comprehensive series was that put out by the
Shanghai Arts and Science Press in 1963. It contained
seventeen volumes in physics, chemistry, and mathematics.
Of this number, nine are in mathematics: four in algebra, two in plane geometry, and one each in trigonometry,

---

[81]C. S. Yau, Analysis of Correspondence Teaching
of Mathematics in the Soviet Union for Forty Years
(Shanghai: Science Publishing Press, 1965).

solid geometry, and analytic geometry.[82] Earlier works
of the same nature included Basic Knowledge on Arithmetic
for Self-Study, by Shen Chao, 1958, of the Natural
Science Series and the Commercial Press, and Factory
Arithmetic and Algebra, by Kuo Sho-to, 1953. Up until
1963, self-study mathematics materials appear to be con-
spicuously free of political content, but it is not
likely that this phenomenon has been maintained through
the reforms of the Cultural Revolution.

Professional Journals

Professional journals in the People's Republic of
China originate from several sources--the research institutes of Academica Sinica, universities and normal
schools, and local professional societies and clubs.
Those intended for teachers are usually published by normal schools and distributed within a limited geographical
area. Several journals devoted to mathematics education,
and at a level of rigor intermediate between The Mathematics Teacher and The American Mathematical Monthly,[83]

----

[82]Titles reviewed included: Yao Chien-ch'u and
Yu Yi-tze, Self Teaching Trigonometry, 1963; Tai Shu,
Self Teaching Algebra, Vol. II, 1964; Yang Yung-hsiang,
Self Teaching Plane Geometry, Vols. I and II, 1964.

[83]The Mathematics Teacher (Washington, D. C.:
National Council of Teachers of Mathematics, Monthly);
The American Mathematical Monthly (Washington, D. C.:
The Mathematical Association of America).

were in circulation in the People's Republic before 1966.
They included:

Zhongxue Shuxue [Middle School Mathematics], Mathematics Department, Wuhan Normal College.

Shuxye Tungxyn [Mathematics Bulletin], Chinese Mathematics Association, Wu-han Section.

Shuxue Jiaoxue [Teaching of Mathematics], Wa tung Normal College.

Shuxue Tongbao [Mathematics Bulletin], Chinese Mathematical Association, Peking.

Chung-hsüeh Shu-hsüeh [Middle School Mathematics], Huanan Pedagogical Institute.

Shu-hsüeh Chiao-hsüeh [Teaching of Mathematics], East China Normal University, Shanghai.

Shu-hsüeh Chiao-hsüeh T'ung-hsun [Mathematics Teaching Bulletin].

Shu-hsüeh Chiao-hsüeh Yüeh-k'an [Monthly Journal on Teaching of Mathematics].

With the exception of Shuxue Tongbao, these journals

appeared to flourish between the years 1956-1959. Most

were patterned after Soviet counterparts and perhaps even

edited by Soviet advisors. Thus when the advisors

returned to Russia, their journals became defunct; or

possibly this spurt of mathematical journalism was a

manifestation of the charged atmosphere of the Great Leap

Forward's increased production quotas.

Zhongxue Shuxue was devoted to the general improve-

ment of middle school mathematics. Its articles con-

sidered specific teaching difficulties, enrichment mate-

rials, the political aspects of mathematics teaching, and

information on Soviet teaching practices. The last
feature was particularly dominant. Reviews of Soviet
textbooks and complete translations of Soviet articles
on mathematics education were included in many issues of
the journal. This practice reached an extreme in July
of 1958 when the whole issue of Zhongxue Shuxue was
comprised of Russian articles.

Less of a vehicle for Soviet influence and more a
forum for readers was the monthly Shuxue Tungxyn, published for senior middle school and beginning college
students and teachers. Many of the journal's contri-
buters were students. Material considered was of a
classical nature and included problems in vector algebra, analytic geometry, calculus, linear algebra, and
probability. Articles of general interest could center
on such topics as events in the history of mathematics
or mathematical games. A feature column "Small Knowledge"
would pose questions to test the reader's knowledge.
Shuxue Tungxyn appeared neutral in its political outlook
and in two expositions concerning digital computer theory
and developments, the machines referred to were produced
by International Business Machines, a monolith among
capitalist enterprises![84]

---

[84]"How a Computer Works," Shuxue Tungxyn (June,
1958), pp. 6-8; "Computer Developments," Shuxye Tungxyn
(December, 1957).

In severe contrast to the nonpolitical nature of
Shuxue Tungxyn, Shuxue Jeaoxue espoused Party policies
to the fullest extent during the period of the Great
Leap. This effort even extended to the publishing of
problems concerning the design and construction of
smelting furnaces during the student steel-making campaigns during the late fifties.[85] Still much of the
journal's contents were devoted to the improvement of
middle school mathematics teaching practices.

The most enduring of all mathematics education
journals was Shuxue Tongbao, an official publication of
the Chinese Mathematical Society in Peking. Production
of this monthly began in November, 1951, and continued
through August, 1966. Due to the prohibitions of the
Great Cultural Revolution, the flow of Chinese professional journals to the West ceased. Nothing is known of
Shuxue Tongbao after this time. The journal has served
as the vanguard for official Party pronouncements
involving mathematics education. Thus the initiation
of the mathematical competitions was first publicly
announced through the medium of this journal, as were
curriculum revisions, adoptions of new official teaching outlines, and texts. During the 1959-1961 period
of debate on national mathematics teaching reforms, its

---

[85]Shuxue Jiaoxue (April, 1959), pp. 27-28.

pages served as a central forum for the exchange of opinions. Later, as the tide of proletarianism swept education into the reforms of the Great Cultural Revolution,
the course for mathematics teaching was indicated by
articles in Shuxue Tongbao. Its contents broadly encompass middle school mathematics education, both on the
formal and informal spare-time levels and also in-service
and pre-service teacher training. With an estimated
monthly circulation of 10,000 copies, its reading audience
is large for a Chinese periodical.[86] Occasionally, a
leading mathematician contributes an exposition urging
reforms, extolling Party educational policies, or presenting some interesting aspect of mathematics; but most often
articles are authored by individual teachers or collective
mathematics teaching and research groups. Frequently,
teachers' contributions would bear the titles, "How I . . ."
or "How to . . .," and go on to present their personal
heuristic approach to a specific teaching situation.
Articles of this nature were very narrow in scope, typi-
fied by the exposition "How to Teach the First and Second
Chapters of Junior Middle School Algebra" in the journal's
February 1960 issue.[87]

---

[86]Raymond Nunn, Publishing in Mainland China
(Cambridge, Mass.: M.I.T. Press, 1966), p. 35.

[87]Wang Yu-ch'ing, "How to Teach the First and
Second Chapters of Junior School Algebra," Shuxue Tongbao
(February, 1960), pp. 17-21.

Over the years, the contents of Shuxue Tongbao
have served as indicators of the general direction the
emphasis in mathematics teaching would take. Material
in the first issues basically puts forth the government's
policies concerning teaching outlines, texts, and teacher
training. Gradually, the journal became more of a forum
for the exchange of ideas between teachers. During the
Great Leap, its articles proclaimed the successes of
numerous experimental teaching projects designed to
accelerate mathematics learning, and advised readers on
conducting spare-time and part-time teaching programs.
By 1960, this enthusiasm was transformed into pronouncements of alarm concerning the decrease in the quality of
student mathematical performance. Then came the debate
on reform. As the school systems once again strove for
academic excellence in the early sixties, the articles
in Shuxue Tongbao became more mathematical in nature.
Seemingly out of place in a journal intended for middle
school teachers, topics such as "Geometric Explanation
of Abnormal Phenomena in Rotation of Hyperbola and the
Special Theory of Relativity"[88] were considered. Translations of articles from Soviet, French, Japanese, and

---

[88]V. G. Liehmulu, "Geometric Explanation of
Abnormal Phenomena in Rotation of Hyperbola and Special
Theory of Relativity," Shuxue Tongbao (March, 1965),
pp. 39-42; translated from Mathematics Teaching Journal,
U.S.S.R. (May, 1965).

American mathematical journals were now finding their
way into the pages of this monthly. Among American contributions appearing in print for the benefit of Chinese
teachers were "The Meaning of the Spectrum Theorem," by
Paul Halmos, and Edwin Hewitt's "Application of Connectivity in Analysis."[89] This brief period of rigorous
mathematical journalism gave way in 1965 to proclamations
of proletarian mathematics teaching. The journal's
presentations now concentrated on Mao's teachings, the
expansion of spare-time mathematics programs, and applied
problems--the Cultural Revolution had engulfed Shuxue
Tongbao.

Student Enrichment Materials

Student enrichment materials in the form of booklets relating interesting or unusual aspects of mathematics are plentiful in Chinese book stalls. Often the
contents of these works are used to impart to the reader
a nationalistic pride in China's past mathematical accomplishments. Typical titles for publications of this
nature include The Study of Geometry by Chinese Mathematicians, Achievements of Chinese Mathematics, and The

---

[89]P. R. Halmos, "The Meaning of the Spectrum
Theorem," Shuxue Tongbao (October, 1964), pp. 47-49;
Edwin Hewitt, "Application of Connectivity in Analysis,"
Shuxue Tongbao (March, 1965); translated from American
Mathematical Monthly (June, 1960).

History of Chinese Arithmetic.[90] Another recurrent theme
featured concerns peasant successes in unraveling the
"mysteries" of mathematics. Wang Li-kêng's Arithmetic
Stories[91] narrates the author's experiences in learning
elementary mathematics, and in doing so encourages others
to attempt the same feat. A story attesting to the
inventive ability of the working class is given in Tuan
Sia-fu's A Method of Measurement,[92] which tells how a
peasant with only three years schooling conceived of a
new method of measurement beneficial in frontier work.
While revolutionary and proletarian themes are reflected
in some booklets, the majority concern enrichment topics
similar to those found in their Western counterparts.

Peasant Handbooks

Books not solely intended for mathematics instruction, but effective in their inclusion and dissemination
of basic mathematics materials are handbooks designed for
peasants. These publications, similar to a "Farmer's

---

[90]Shiu Shen-fong, The Study of Geometry by Chinese
Mathematicians (Peking: China National Science and
Technological Association, 1956); Ten Tung-chi, The
History of Chinese Arithmetic (Peking: The People's
Education Press, 1957).

[91]Wang Li-kêng, Arithmetic Stories (Peking:
Shing-Hwa Book Store, 1958).

[92]Yu Chun-hsien, A Method of Measurement (Shanghai:
Commercial Press, 1952).

Almanac," contain many sections on topics considered
relevant to peasant life, such as animal husbandry,
first aid, letter writing, horticulture, and simple mathematical calculations. An examination of the mathematical
contents of these books provides an accurate picture
concerning the type and level of mathematics taught in
peasant spare-time schools. The Farm Village Practical
Handbook,[93] published in 1966, supplies such a picture.
Thirty pages of its 504 are devoted to mathematical
calculations. Specific topics covered in these pages
include:

1. Use of the abacus
   Practice involving the four operations.

2. Use of mensuration formulae
   Computation of area for: squares, rec-
   tangles, trapezoids, triangles, and
   circles.
   Method of dissection used on irregular
   convex polygons.
   Computation of volumes of: cubes, rectangular prisms, cylinders, cones,
   pyramids, and frustra.

3. Estimation
   Estimation of rice yield per mou of land.
   Estimation of supply of mulberry leaves
   available during silkworm season.

4. Simple percentage problems

Presentations are straight-forward and provide little
pedagogical motivation. The problem-solution style is

---

[93]Farm Village Practical Handbook (Shanghai:
Shanghai Publishing Company, 1966).

reminiscent of <u>Chiu-chang Suan-shu</u>, the two thousand-year-old Chinese mathematical classic.

Effectiveness of Publications

Popular Chinese mathematical publications such as these have supplied a vital link in transmitting mathematical knowledge to a wide audience. Although short-lived due to their cheap construction, the booklets' quality is sufficient to permit them to serve as valuable and readily accessible learning tools. Their low price has permitted many schools to acquire libraries of such booklets. This method of bringing mathematics to the masses could well be adopted by other developing nations of the world.

Political Indoctrination: A Facet of Mathematics Instruction

In the People's Republic of China, all school content, including mathematics, serves as a media for advancing the government's political policies. Many Chinese school subjects have always been nationalistically charged and anti-foreign, but formerly these subjects could almost always have been categorized as being in the fields of social science or literature.[94] Under

---

[94]For a detailed discussion of this educational policy see Victor Purcell's <u>Problems of Chinese Education</u> (London: Kegan, Paul and Trench, Trubner and Co., 1936).

the Communists this policy has been expanded to include
every subject taught. Education, indoctrination, and
propaganda are synonymous in the thinking of China's
leaders.

> Who is a propagandist? Not only is the teacher
> a propagandist, the newspaper reporter [is] a
> propagandist, the literary writer [is] a propa-
> gandist, but all our cadre in all kinds of work
> are also propagandists . . .[95]

Universally, education has always been the primary means
employed by nations to indoctrinate their young in
loyalty to the state. Propaganda is a relative term--
what is propaganda to an observer may be deemed reality
by its promoter. Even in the light of these two acknowledgments, the Chinese experience presents an extreme
extension of both concepts.[96] To a Westerner, Chinese
indoctrination is absolute, but to a Chinese, heir of
the Confucian tradition, ideological training in education has always been a vital part of schooling and is
expected. Perhaps a more acceptable synonym for ideological indoctrination in the Chinese context would be
"nationalistic morality." This new morality of post-1949
China is based on the "five loves" of Article 42 of the

---

[95]Mao Tsê-tung, "Opposing Party Formalism"(February, 1942).

[96]Theodore Chen, "Education and Propaganda in
Communist China," The Annals of the American Academy,
American Academy of Political and Social Science, Vol.
227 (September, 1951), pp. 135-145.

Common Program: love of motherland, love of labor, love

of science, love of the people, and protection of public

property. After liberation, the "five loves" became a

popular slogan emphasizing the socialist spirit of the

new state. In response to this slogan, various Chinese

educators produced manuals and guidebooks in which they

advocated the correct techniques for teaching patriotism

and instilling the five loves in pupils' minds. The

teacher's and school's task in this movement is clarified

in a work by Hu Yen-li:

> . . . the implementation of Five Loves education
> is not the responsibility of the political
> information teacher only, but it is the common
> task of the entire body of teachers. Moreover,
> this is not a form of education that should be
> carried out only in political information
> classes. Rather, education should be carried
> out in all subjects, in all classes, and in all
> extracurricular activities, with the Five Loves
> as its core.[97]

Similarly, an editorial in the March, 1955, issue of Jen-

min Chiao-yu, a monthly journal published by the Ministry

of Education entitled "How to Strengthen Moral Education"

advised that moral education be guided by Marxist princi-

ples. Teachers should aim to inculcate a Communist world

outlook and to impart to the student a will to fight for

socialism.

---

[97]Hu Yen-li, How to Carry Out Five Loves Education
(Peking, 1951), p. 2.

Exhortations such as this were readily followed by teachers and writing groups composing classroom texts and materials. Often in times of educational flux, as during periods of policy change, the writing of these groups replaced the regular government-issued textbook series. In attempts to make these works politically relevant, issues concerning China's world position, her allies, and her enemies were selected from newspapers and incorporated into the teaching materials. Hu indicates how such a policy was employed by the Nanking Normal College's elementary program.[98] The most relevant political issue of the time was the Korean War. During the winter of 1950 and the spring of 1951, an extensive campaign directed at opposing American "aggression" in Korea was conducted throughout China. A central agency for arousing public resentment was the school within which special "learning sessions" on the Korean situation were taught. In arithmetic, statistics concerning the two factions, "freedom-loving peoples" and "imperialist aggressors"; populations, troop losses, etc. were used in computational schemes designed to illustrate the inferiority of the aggressors. Typical in the Nanking

---

[98]Translated in Dennis Doolin and Charles Ridley, The Genesis of a Model Citizen in Communist China (Washington, D. C.: U. S. Office of Health, Education and Welfare, 1968), p. 34.

school's problem themes were:

1. Calculations concerning the prizes of war
   obtained by the Liberation Army at each
   victory.
2. Calculations of production and construction
   in this city [Nanking] for the past two
   years.
3. Calculations of the losses of the American
   imperialist in its invasion of Korea [sol-
   diers and weapons].
4. Calculations of the strength of the camp of
   world peace [population, area, products and
   military].[99]

Propaganda in Mathematics Instruction

The extent of the amount of political material in
mathematics texts varies according to the sophistication
of the readers and, apparently, the mathematical and
pedagogical competence of the writers. Most extensive
political exposure is provided in primary texts or
peasant handbooks. Secondary texts, prior to the Great
Cultural Revolution, concentrated, for the most part, on
mathematics. These texts did, however, give frequent
reference to China's historical accomplishments in
mathematics. Numerous paperbacks which also focus on
historical mathematical achievements are available in
the bookstalls of China.[100] These phenomena are intended

---

[99] Ibid.

[100] For example: Hua Lo-kêng, Discussions Starting
From π of Tsu Ch'ung and Discussions Starting From Yang
Hui; Yen Tung-chung, Achievements of Chinese Mathematics;
Shui Shen-fung, The Study of Geometry by Chinese Mathema-
ticians.

to foster a nationalistic pride in China's early mathematical advances, as well as a resentment of the lack of recognition of this fact in the West. Books written by professional educators are less likely to be as overtly political as those written by dedicated cadre. In the aftermath of the Great Cultural Revolution, all school texts are being rewritten by people's committees to cleanse them of "bourgeois contamination." As a result of this "cleansing," all textual material, including mathematics, will be further burdened with political rhetoric.

Journals intended for mathematics teachers are constantly reminding their readers of the importance of promoting correct political attitudes through teaching.[101] Some journal authors venture to offer examples of politically flavored mathematical problems:

> An American ship that is known to be 120 pu in length is observed from the shore by one of our soldiers. If for his gaze to traverse the length of the ship, his eyes must move .0072 radians, has the ship violated China's twelve mile limit?

> The population of the U.S.A. taken in February 1958 was 170 million people. If 5,200,000 are unemployed, what is the proportion of unemployed to employed people in America?

> American cotton production for 1958 is 5/7 x 20,000,000 tan [133.3 pounds]. If we produce 34,000,000 tan of cotton more than our present

---

[101]Chung-hsüeh Shu-hsüeh, Huanan Pedagogical Institute; Shuxue Tongbao (Peking: Chinese Mathematical Society).

output, we would double that of the U.S. What
is our present rate of cotton production?

The U.S. steel output for 1958 was 5,900 million
tons, this was 2,816 million tons less than
1957's output. If steel production keeps reducing at this rate, what will the U.S.'s steel
output be in 1960?[102]

Problems such as these find their way into regular and

spare-time middle school texts and instruction. Thus

captive audiences are presented a distorted picture of

the outside world. Militarism is a recurring theme

paralleling school instruction in the Yenan period.

To liberate Formosa, the liberation aircraft
searching in the air is 12 miles away from the
liberation gunboat and 16 miles from the American Imperialist fleet, the angle A is 60°
[referring to given diagram], what is the
distance between the gunboat and the fleet?[103]

Even very young children in nursery school are not spared

from this form of teaching, although the problem situa-

tions may be made more concrete. Foreign visitors to the

nursery school of Peking Number Three Textile Mill

reported the teacher citing "The United States imperial-

ists have sent 10 planes to bomb Vietnam," as she placed

---

[102]First Middle School Research Group, Canton
Province, Ching-min city, "Follow Party Rules to Revise
Mathematics," Chung-hsüeh Shu-hsüeh (February, 1959),
pp. 19-20; Mathematics Section, Second Middle School,
Canton, "Combine Mathematics Teaching with the Present
Situation," Chung-hsüeh Shu-hsüeh (March, 1959), pp. 1-2.

[103]Yeh-wei, Slave Education of the Middle School
in Communist China (Hong Kong: Freedom Press, 1951), p.
88.

ten airplane cutouts on the board. She then took out a
toy rifle, aimed, and fired. Nine airplanes were removed.
"Our heroic Vietnamese brothers have shot down nine
planes, how many are left?" This scenario was staged for
an audience of five- and six-year-olds![104]

The mathematics problems of the Fourth Spare-time
School for Workers and Staff Members in Peking indicate
the strong dependence on newspaper-supplied statistics:

> In the United States of America, the number of
> half-starved people is twice the number of
> unemployed, and is five million less than the
> number of people who live in slums. As one-
> half the number of slum-dwellers is eleven and
> a half million, what is the number of unemployed
> in the United States?
>
> The total number of Japanese policemen and police
> reserve corps armed and equipped by the U.S.A.
> is 218,000. It is known that the number of
> policemen is 68,000 more than the number of
> police reserve corps, how many policemen and how
> many police reserve corps have been armed and
> equipped by the U.S.A.?
>
> Thirty-one U.S. B-29 bombers and over forty U.S.
> fighter planes invaded our sky over Antung on
> April 12. These bombers aimlessly bombed the
> city of Antung. The number of bombs dropped by
> these planes was eighty more than the bombs
> dropped by the twelve U.S. B-29's on March 31st
> into the city of Linkiang. The total number of
> bombs dropped during the two invasions was 140.
> How many were dropped in each invasion?[105]

[104]Norman Webster, "China's Factories Mind the
Children," The New York Times (November 19, 1970), p.
56.

[105]Chi Tung-wei, Education for the Proletariat in
Communist China (Hong Kong: Union Research Institute,
1956), p. 46.

Each problem is a vignette of indoctrination from which
a lengthy discussion will result. In many problems, such
as the last, the mathematics principle being communicated
is all but obscured by superfluous information. One can
picture the astute student questioning the economy of
such an air raid! Repeated problem situations used to
reinforce the "five loves" are (1) past exploitation of
the masses, (2) national accomplishments, and (3) aggres-
sive actions of the "enemies of the people." While these
themes remain constant, their particular emphasis may
shift with the current political situation, warranting
a re-editing of text materials. When Sino-Soviet friend-
ship was blossoming, mathematics problems reflected good-
will towards and a desire to emulate China's socialist
brother. Since the Sino-Soviet rift, the situation has
reversed itself, and Russia now shares the brunt of
moralist mathematical problems.

In manifesting the glories of China's liberation,
situations are frequently related from the Chinese past
to illustrate the exploitation that existed at the time.

In the old society, there was a starving family
who had to borrow five tou [about 200 pounds]
of corn from a landlord. The family repaid the
landlord three years later. The greedy landlord
demanded 50% interest compounded annually. How
much corn did the landlord demand at the end of
the third year?[106]

---

[106]"China's New Math and Old Problems," The New
York Times (March 9, 1969), p. 18.

The solution of this problem reveals the family had to pay nearly seventeen tou of corn to settle its debt. A continued discussion and more problems depict the family losing its land to the landlord who in turn rents the land back to it. The moralistic lesson cited is that in the old society the landlords manipulated the counters of their abacuses in exploitation of the working class. Now in the new society, students must learn to move them to serve the peasants.[107]

Under the fervor of the Cultural Revolution, even problems such as the above had to be revised to be more proletarian in their presentation. A report from Middle School Number Twenty-three in Tiensin illustrates how this revision was accomplished.[108] A certain problem presented in a first-year mathematics class required the students to calculate how much a poor peasant family in the old society had to pay in rent, repayment of debts, and taxes in terms of grain at the year's end. No consideration was given as to how much grain was left for the poor family! The students pointed out that this was not an account of "blood and tears" of the poor peasant,

---

[107]Ibid.

[108]"Persevere in and Intensify the Educational Revolution," Kuang-ming Jih-pao (Peking, March 20, 1970, translated in S.C.M.P. 4631 (April 8, 1970), pp. 60-71.

but an account of exploitation by the landlord. Further
they emphasized, "Such a problem is designed to train us
as accountants for landlords and rich men." As a result,
the teachers voluntarily re-examined their class feelings
and together with their students examined the situation
in depth to discover that only thirty-six catties of
grain were left for the peasant family. The landlord
received 1,764 catties.

An example that truly brought proletarian politics to the fore for the Tiensin middle school students
was the story of worker Hua:

> Master worker Hua, an old worker of Number Three
> Rug Manufactory of Tiensin, used to weave carpets
> for the capitalists in the old society. He earned
> only 6 yuan a month, from which 20 fen would be
> deducted for meals if he did not work for a day.
> When he was only fourteen [years old], Master
> Worker Hua was ill and did not work for 25 days
> in April. A total of 5 yuan was deducted from
> his wages for meals. He also borrowed 5 yuan from
> the management of the factory for medicine. As a
> result, Master Worker Hua did not receive any
> wages for that month, but instead owed the capi-
> talist 3 yuan, which was a negative amount.[109]

The class content of this lesson was pointed out as being
exceptional, as it moved the students to tears.[110]
Unfortunately, the algebraic principle involved, negative
quantities, is all but obscured in the presentation. A

---

[109]Ibid., p. 63.

[110]A figurative expression indicating the students
were deeply moved.

more interesting example of negative numbers is presented
to Chinese students through a consideration of the United
States national debt. It is pointed out how this deficit
of hundreds of billions of dollars was accumulated as a
result of suppression at home and imperialist aggression
abroad.[111]

Indoctrination, a Hindrance to Mathematics Learning

Problems such as these certainly satisfy the
dictates that politics "take command." While seldom
preaching the "five loves" directly, the implications
concerning the outrages of capitalism or imperialistic
aggression instill a reverence and protective attitude
for the motherland. Since all school subjects are
impregnated with ideological materials, mathematics must
also lend its support to this effort in order to preserve
teaching continuity. As noted in two of the examples
illustrated above, the prescription for indoctrination
may lessen the mathematical teaching efficiency of a
problem situation. Excessive reading materials, including statistics in problems, cause an additional burden
on the pupil. If the pupil's reading ability is insufficient, he may never come to realize what mathematical

---

[111]"Put Mao Tse-tung Thought in Command of Cultural Courses," Red Flag (Peking, March, 1971).

task is implied. Teachers and students from the People's Republic of China indicate that one particular area of difficulty that secondary students of mathematics encounter is word problems.[112] The stress on political material in problem situations could very likely be a contributing factor to this phenomenon.

Surveying Practice: A Reinforcement of Geometry Instruction

To a Western observer, perhaps the most unusual feature of the modern Chinese middle school program is the inclusion of surveying practice in general mathematics studies. Such a course, while unwarranted in the secondary schools of highly industrialized nations, is certainly in accord with China's basic developmental needs. Land survey is not a recent addition in Chinese middle school mathematics. Early twentieth century missionaries had incorporated survey work in their instruction of trigonometry,[113] but with the formalization of school curricula, this practice was eliminated. In their attempts to orient mathematics towards the reconstruction needs of the nation, starting in 1957, the Communists

---

[112]Contacted through private correspondence.

[113]H. Hermann, "On the Surveying Instruments Described in Parker's Trigonometry," Educational Review (August, 1910), 3: 1-3.

reinstituted this subject in middle schools. While some
schools accommodated survey work during class time, many
designated it as an extracurricular activity. The techniques studied were of the most elementary nature and
primarily concerned mensuration and layout. Field procedures relied heavily on geometric principles rather
than numerical calculations. Despite the basic level of
student experience, it still provided a valuable foundation for agricultural construction and possible future
military participation.

Surveying Activities

The level of difficulty of the surveying exercises
vary according to grade status.[114] Thus a seventh-grade
first-semester class might be required to construct a
meter measure and field "compass."[115] The bamboo compass
would be calibrated in its span to either one, two, or
one-and-a-half meters. Students would then be required
to lay out a straight line on the school grounds and
measure it. Four methods of measurement are used--
visual estimation, pacing-off, employment of a field
compass, and improvised surveyor's tape. Standard

---

[114]Tsi Pei-ting and Chan Tse-min, "Middle School
Survey Practice," Shuxue Tongbao (May, 1957), pp. 32-37.

[115]The term compass denotes a divider-like instrument used for "walking-off" distances.

distances of ten, fifty, and one hundred meters are established within the school's ground to provide students with a comparison for visual distance estimation. Each student's pace is calibrated in meters for his personal knowledge and future reference. In their second semester, students acquire practice with a surveying square in laying out right angles and measuring the height of given objects. Calculation of the area of a plot of land whose boundaries form an acute triangle is also required. Special school-related projects, such as laying out a basketball court or school garden, could be assigned as required by circumstances.

First term eighth-grade work would include the layout of a straight line for some distance over both level and uneven terrain. Measurement and layout of various angles using a protractor is required. The construction of a perpendicular to a given line is also undertaken, as well as techniques for surveying across inaccessible regions, such as a river or around a mountain. Second-term experiences include further practice in surveying to inaccessible reference points and methods for measuring height. Attention is now also given to a consideration of projective techniques for reducing field data to drawing board scale, and the construction of plans from observations.

Ninth-grade tasks might include the construction of a line parallel to a given reference line, construction of straight lines through obstacles, measurement of distance through obstacles, the layout and calculation of area for various polygonal plots, and the measurement of elevation-employing levels. At this stage of activity much of the work would also involve drafting practice.

Although actual class time expended in surveying practice is small, three per cent,[116] and field practice is often replaced by classroom discussion, the benefits to the state of students' experience are substantial. By being exposed to this sequence of activities, a middle school graduate will be better prepared for commune settlement, frontier construction, and military field maneuvers.

The influence of the Cultural Revolution further removed middle school mathematics studies from the realm of theoretical subjects and assisted in making them mere tools for production and, indirectly, indoctrination. Whether this practice will sociologically and technically strengthen the People's Republic of China remains to be seen.

---

[116]Hsu Chien-hua, "My Views on the Surveying Practice for Middle School Students in Studying Mathematics," Shuxue Tongbao (November, 1965), pp. 20-21.

Conclusions

During their twenty years of educational dominance in China, the Communists have exhibited an extraordinary flexibility in their school policies and pedagogical techniques. The resulting academic reforms have always been in response to changing national goals—both political and industrial—but have also reflected a search to find the best fit between education and contemporary social and economic needs. This search has been especially obvious in the numerous mathematics education innovations that have taken place during this period.

By approaching the problem of upgrading the quality and popularity of mathematics studies by many diverse means, Chinese educators have demonstrated a firm commitment towards achieving high standards and widespread effects of mathematics instruction. If this determination persists, it appears to be merely a matter of time before these goals will be obtained.

CHAPTER VI

SUMMARY AND CONCLUSIONS

The Historical Evolution of the Present Model of Mathematics Education

A Century of Continuous Reforms: 1870-1970

Nineteenth century China's initiation into the world of nations was abrupt and traumatic. With the large-scale influx of foreigners came a series of disastrous wars that shook the Chinese from their intellectual complacency. The Mandarins realized that the defense of their empire rested in securing a modern technology and through it a modern military force. Western knowledge, primarily languages and sciences, was sought as the foundation for industrial development. Among the sciences, mathematics was considered the most important, and its study was pursued at all government translation and arsenal schools. This action was instituted over the objections of court conservatives. Finally, with the inclusion of mathematical questions on the civil service examination, the discipline was designated worthy of scholarly pursuit and its position secured in future Chinese school curricula.

The Manchu government's establishment of a modern national school system was based upon Japanese educational exemplars. Partially, this appeal rested upon the Japanese disciplined approach to instruction which fostered a subservience to authority and a respect for cultural and ethical traditions. The new curricula reflected traditional thinking by stressing humanistic and ethical subjects at the expense of scientific ones. Despite this strong conservative bias, a modern structure for mathematics education was established. The study of arithmetic, algebra, geometry, and trigonometry was prescribed for all students. Mathematics teaching emphasized calculation and promoted memorization, two learning characteristics prominent under traditional schooling.

Predominance of American influence in the Republican and Kuomintang periods brought with it a progressive educational philosophy centering, in principle, on the development of the individual. Although limited mathematics curricula experimentation took place, most mathematics teaching retained the pedagogical characteristics of the previous period. At the secondary level of instruction, the study of algebra focused on the mathematical concept of function. This paralleled the vogue of the time in American teaching. The study of analytic geometry, adopted from American college offerings, was added at the senior middle school level. Geometry teaching had developed into a required display of graphic proficiency reminiscent of the calligraphic standards set for a traditional scholar. An American-inspired adoption of educational testing and measurement procedures resulted in the publication and use of standardized mathematics tests. By the 1930's, through this testing, Chinese education had reverted to a system of qualifying examinations similar in their designs and purpose to the traditional civil service tests. A student's future educational and occupational success was made contingent upon the passing of a series of state-administered examinations.

The Communist proletarian approach to education did little initially to offset the impediments to creative mathematics learning inherited from China's past. Under the People's Government, mathematics education became more universal and extremely utilitarian in nature. Instruction was combined with political indoctrination, with the thoughts of Chairman Mao replacing the doctrines of Confucius.

During the Great Leap Forward, Soviet influence combined with the urgency of achieving rapid industrialization, resulting in the enlargement of mathematics studies at all levels of the educational system. This emphasis was strongly coordinated by government-issued teaching outlines and the cooperative efforts of mathematics teaching groups. In the search to achieve the utmost efficiency in mathematics instruction, psychological research into the processes of mathematics learning became a state-sponsored activity. Competitions patterned after the Soviet Mathematical Olympiads were used to locate "budding" mathematical talent, recalling the Mandrins' scheme of supplying scholars for the Chinese state. The tendency of traditional attitudes and practices to stymie full-scale mathematical innovation continued, up until 1966, when the Great Cultural Revolution swept the Chinese educational scene.

Official government pronouncements urging reforms in all aspects of Chinese life provided the impetus for dramatic educational change. Under these directives the actions of the Red Guards and revolutionary school committees have resulted in the elimination of examinations, abbreviation of curricula, and abolition of formalism in teaching. Mathematical studies were tailored to the agrarian and production needs of specific geographical locales. Thus, contemporary Chinese mathematics teaching reflects extreme utilitarianism based on the social needs of the state, a phenomenon reminiscent of the Chiu-chang Suan-shu.[1] Once again, the abstract nature of mathematics is being neglected. Even in the research institute of Academica Sinica, little pure mathematical research is

---

[1]See page 13.

conducted; investigations are usually narrow in scope and directed to some specific problem of application. The Chinese are again looking inward educationally and reasserting their self-identity. This reassertion has resulted in a new and interesting activity-oriented approach to mathematics instruction, an approach that may prove informative to educators outside of the Middle Kingdom.

Cultural Impediments to Effective Mathematics Innovation

A society with thousands of years of uninterrupted history accumulates social traditions and prejudices that do not easily give way to innovation. During the modern period, under consideration in this study, the effectiveness of mathematics teaching in Chinese schools has suffered greatly from inherited attitudes.

Preservation of one's social image or "face" is still very much a part of Asian life. A heuristic classroom approach is particularly conducive to mathematics teaching. In the Chinese situation, children were often afraid to ask or answer questions, lest they appear ignorant before their peers and thereby bring disgrace to themselves and their families. In turn, teachers discourage questions in order to preserve "face" and insure an unchallenged position of authority. Unfortunately, due to the inadequate training of mathematics teachers,

many are simply unable to answer student questions. This
tendency of students and teachers to avoid discussions of
mathematical topics diminishes the spirit in which productive mathematics learning may best be achieved--one of
open inquiry and experimentation. The result has been a
strong dependence on formalization and rote memorization
in mathematics education.

Until recently, Chinese middle school curricula
have contained an overburdening variety of subjects. As
a result of such scheduling, the amount of time a student
could devote to mathematical studies has been limited.
This practice is due in part to the urgency with which
China seeks to enter the modern world through education;
but it also reflects the enduring ethnocentric conception
that "Western knowledge" is rather elementary in nature
and can be easily mastered by Chinese schoolboys. One of
the results of the Great Cultural Revolution is the
abbreviation of all school curricula and the apparent
elimination of this problem.

Throughout most of the twentieth century, the
tradition that scholars should abstain from any activity
that pertains to manual labor has prevailed. Thus,
mathematics studies have often excluded from consideration problems dealing with the applied aspects of the
discipline. Yet this fact has not brought about greater
concentration on the abstract aspects of mathematics.

Indeed, studies have often concentrated on neither approach, but rather on the acquisition of a collection of classical problem-solving techniques. The structure of mathematics and the interrelationship of its various elementary branches are obscured. The Communists are presently attempting to stress social applications of mathematics. This emphasis does not develop mathematics as a self-contained body of knowledge useful for itself, with diverse and far-reaching applications.

While these specific problems are peculiar to the Chinese educational scene, they spring from basic issues relevant to every society: "What is the position of mathematics in relation to the self-realization of individuals?" and "What is the function of mathematics in respect to the development of a society?" As long as Chinese educators do not resolve these questions, mathematics education will not achieve its full potential in China.

A Comparison of Mathematics Education in China During the Years 1925-1949 and 1949-1970

A qualitative assessment of the advances made in Chinese mathematics education requires comparative norms. To seek these norms outside of China and Chinese society would be almost meaningless; therefore, an internal comparison should be sought. An examination of modern Chinese

295

history reveals two periods in which a national identity was secured and viable institutions and reforms established. These periods--the Nationalist, 1925-1949, and the Communist, 1949-1970--readily lend themselves to comparison.

Both the Nationalists and the Communists based their educational systems on admired foreign models. The Nationalists adopted their educational philosophies and organization from the United States. The Communists chose a Soviet socialist exemplar. Both imitations were complete in almost every aspect, with teaching methods, curricula, and textbooks borrowed from the respective countries. In both instances, these policies resulted in an alienation of students from the realities of their society. Curricula were rigorous and classical, encouraging formalism and memorization. Each government depended on a system of selective examinations for filtering students into higher education. Education became an elitist activity, producing intellectual stratification reminiscent of traditional China. In the most recent period of the Communist era, the situation has been rectified by the reforms of the Great Cultural Revolution. Examinations have been abolished and curricula abbreviated and tailored to reflect socioeconomic needs of Chinese society.

These reforms would satisfy previous criticisms
directed at Chinese mathematics education. The elitist
nature of mathematics teaching denounced by Pêng Chun
Chang has been eliminated.[2] Indeed, Chang's proposed
objectives for mathematics studies--preparing students
for exploring and pioneering, new community building,
and scientific production and organization--appear very
evident in the present Communist education system. The
League of Nations critique of the Kuomintang's system of
mathematics education centered on its formal nature and
advised a reorientation of curriculum around village life
and agricultural and practical needs.[3] This reorientation has been accomplished. A later commentator on this
same system was Jen-chi Chang, who strongly advocated a
decentralization of education, emphasis on locally run
rural schooling, and the establishment of a new educational philosophy founded on the socioeconomic needs of
the rural population.[4] In directing his comments to
mathematics instruction, Chang urged the teaching of
simple bookkeeping procedures in school to assist farmers
in drawing accounts.[5] The Communists have carried out

---

[2]See page 90 above.

[3]See page 96 above.

[4]Jen-chi Chang, <u>Pre-Communist China's Rural School
and Community</u> (Boston: The Christopher Publishing House,
1960).
[5]<u>Ibid</u>., p. 101.

both the League's and Chang's proposals in their restructuring of the Chinese educational system.

Qualitatively and quantitatively, Communist
advances in mathematics education have far exceeded those
of the Nationalists. This progress is a result of the
state of relative peace that China has experienced in the
last twenty years, allowing the establishment of reforms,
and the government's designation of mathematics as a
priority academic subject.

Upon seizing power, the Communists established a
strong centralized control of education. The Ministry of
Education issued detailed teaching outlines for all subjects, including mathematics. Mathematics research
groups in individual schools saw to it that these outlines were followed. Textbooks and teaching guides
strictly adhered to official syllabuses. While the
resulting teaching was not imaginative, it was standardized. Nationalist attempts at such standardization
faltered in the confusion of the times.

Promotion of scientific studies, especially
mathematics, in the People's Republic of China is
directed at achieving rapid industrialization. In contrast, Nationalist academic priority seemed to be the
study of English, promoting a close association with the
United States. Psychological research into the varied
aspects of mathematics learning, the wide dissemination

of popular mathematics enrichment materials, establishment of Mathematical Olympiads, and the resulting student mathematics study groups, as well as the organization of extracurricular mathematics activities, all did much to improve the climate and quality of Communist mathematics instruction.   Instructional time devoted to mathematical studies at the secondary level increased constantly from 1950 onward, until by 1962 it accounted for 18.9 per cent of school time (see Table 23).   This figure compares to the Nationalists' highest figure of 15.3 per cent obtained in 1935.   Although records are only available until 1962, it is probable that in the years between 1962 and 1966, the curriculum of key schools increased this figure to well over 20 per cent.   Thus, due to stable social conditions and national priorities, the quality and scope of mathematics education during the period 1949-1970 far exceeds that realized in any other period of modern Chinese history.

TABLE 23

TABULATION OF PERCENTAGE OF MIDDLE SCHOOL
TIME SPENT IN MATHEMATICS INSTRUCTION
1903-1962

Educational System	Total Subject Hours	Hours Devoted to Mathematics	Percentage of Program Spent in Mathematics
Manchu*			
1909	432	20	4.6

TABLE 23 (Continued)

Educational System	Total Subject Hours	Hours Devoted to Mathematics	Percentage of Program Spent in Math.
Republican**			
1912	129	19	14.5
1921 (Science)	123	22	17.8
(Arts)	121	16	13.2
Kuomintang***			
1928	336	49	14.6
1933	194	24	12.8
1935 (Science)	183	28	15.3
(Arts)	183	22	12
1940	187	26	14
1947	174	21	12
Communist***			
1950	171	22	12.8
1956	191	36	18.8
1962	190	36	18.9

* Five Year Program
** Four Year Program
*** Six Year Program

Chinese Experience as an Educational Model for Other Developing Nations of the World

Despite varied claims of superiority, the Chinese
have never deluded themselves concerning their state of
national development. China, a populous nation with
partially tapped resources, is attempting to achieve
modernity by transforming an agriculturally based economy
to one dependent on heavy industry. As such she can be
classified as a developing nation. The Chinese have used
scientifically oriented education as the main force in
this developmental transformation. China's prolonged

struggle to meet the West on equal terms has been marked by numerous educational experiences. The lessons learned from these experiences could well benefit other young nations beginning down the torturous paths of progress. Two conclusions, in particular, can be distilled from modern Chinese history concerning mathematics education:

1. The mathematics curriculum of a highly industrialized nation cannot be superimposed _in toto_ into the educational system of a traditional society.

2. The enormity of nurturing a modern mathematical outlook and facility in the young of a developing nation requires a variety of psychological and pedagogical approaches.

Chinese experimentation with and eventual rejection of Japanese, American, and Soviet mathematics curricula testifies to the futility of slavish educational imitation. In the three instances, emulation was indeed thorough. Texts and course outlines were adopted in full from the respective countries, and large-scale exchanges of school personnel were carried out. Still, the results of these activities did not produce the intellectual dividends desired. In all cases, the sociological and ideological foundations of the foreign exemplars were alien to the Chinese setting. Mathematics curricula and pedagogy imported into China were based on the legacy of a Western scientific tradition unknown in the Middle Kingdom. The few Chinese students who could assimilate the new mathematical knowledge had difficulty re-interpreting it within the Chinese mileau. Chinese science had been formulated upon socially conceived utilitarian needs through a process of empirical-inductive reasoning. In contrast, Western science has resulted from an industive systemization of the universe. After fifty years of attempting to reconcile Chinese tradition with Western educational practices, Chinese educators evolved a system consistent with their social and economic state of development. Once again utilitarian empiricalism, in the form of solution schemes for agriculturally and industrially oriented mathematics problems, has become the foundation of Chinese mathematics instruction. Western mathematics curricula have been screened and topics of study selected for their relevance to China's immediate needs. These topics are taught with stress on their applied nature, or incorporated into a socio-political analysis of proletarian problem situations. The latter approach reflects the regime's continual emphasis on political indoctrination. This practice continues a traditional Chinese policy of injecting official philosophy, in this case the tenets of Marxist-Leninism as interpreted by Mao into the teaching of an academic discipline.

Developing nations are hindered in their attempts at educational reforms by the magnitude of the task confronting them and by their limited financial capabilities. The difficulties of establishing schools and training teachers is often compounded by the need to convince the population of the merits of education. In the instance of subjects like mathematics, which may be culturally alien to the societies in question, this task becomes one of primary importance. Chinese experience, culminating in the Communists' policy of "walking on two legs," has given rise to several partial solutions for these problems. Adaptation of some of the following educational methods and innovations might well be suited to other developing nations:

1. The Chinese commitment to undertake psychological research involving mathematics learning has helped them adapt foreign mathematics curricula to their own needs. Through observation and experimentation, China's educational researchers have isolated particular mathematics learning and teaching difficulties, and are directing their attention to reducing them. Such research efforts have led to a reculturization of adopted mathematics materials, a necessary step toward the avoidance of educational alienation.

2. The present Chinese proletarian approach to education rests largely on activity-centered learning.

In mathematics teaching, this technique is replacing lecture-dictated theory with environmentally supplied applications of mathematics principles.

An illustration of this process was related by a Western visitor to a Kuangtung Commune.[6] He observed a secondary class receiving a physics-mathematics lesson during the process of repairing a truck engine. In the course of the students' work, the instructor explained the use of calibrated instruments and the calculation of necessary measurements to grind valves and shorten or lengthen the piston stroke. The theory of an internal combustion engine was also considered with the students calculating the areas, volumes, and compression ratio for their particular engine. Upon completion of the project, the students felt rewarded by the satisfaction of contributing to the welfare of their commune by the use of mathematical knowledge. Such a learning scheme, by its concrete nature and the opportunity for student involvement, reduces many of the difficulties formerly encountered in mathematical conceptualization, and supplies immediately appreciated goals.

The Chinese surveying curriculum approaches geometry teaching in the same manner.[7] Geometry has been

---

[6]An American visiting China in 1966. Contacted through private correspondence.

[7]See page 283 above.

liberated from the tediousness of graphic manipulation and memorization and incorporated into the process of nation building. In an agricultural society, the opportunity for students to lay out fields, design irrigation facilities, and construct small structures is far more appealing than the recitation of uncomprehended theorems, as in the past.

3. The distribution of inexpensive, popular, mathematics books, organization of extracurricular mathematics activities, and the institution of mathematics competitions have done much to encourage the study of mathematics among young Chinese. Such devices expose students to the varied recreational, historical, and theoretical aspects of mathematics and supply them with a background otherwise unobtainable in the Chinese situation. The prestige of successful participation in the national mathematics competitions has provided Chinese students with an artificial inducement to strive for mathematics-learning achievement. These innovations, although derived from traditional institutions, have done much to psychologically prepare Chinese students for modern mathematics studies.

4. An enduring problem in Chinese education has been the training of sufficient numbers of teachers to staff the expanding school system. While this problem remains unresolved, the Chinese improvisations directed

towards it are worth examining. In China, the use of television universities has been effective in both training
new mathematics teachers and updating the knowledge of in-
service personnel. Some developing nations have experimented with televised instruction with disappointing
results.[8] Often these attempts were not directed at a
selected population, nor involved specific disciplines;
thus they lost much of their impact. In contrast, the
Chinese experiment concentrates on a very few priority
subjects for a selected audience and supplies a thorough
back-up employing guidance stations. Mathematics research
and teaching groups have provided a strong influence in
the upgrading of Chinese mathematics teaching. Their
collective effort assures standardization of teacher
presentation and provides novice teachers with further
pedagogical training.

The appeal of these innovations should not mislead
as to their applicability to other cultures and countries. If the Chinese experience in mathematics education has shown one thing, it is that curricula and
pedagogy must be relevant to the society in question!

---

[8]I.e., Malaysia.

Conclusions and Predictions

The final analysis of the effectiveness of any
mathematics education reforms rests in considering the
answers to two questions: "What is the position of mathematics education in advancing national goals?" and "Is
this position being fulfilled?" Since the introduction
of modern mathematical studies into China during the
Manchu Dynasty, proficiency in the discipline has been
considered necessary for advancing the State's industrial
and military goals. This policy was very much in evidence
during the early administration of China's present
government. Specific time quotas were established for
overtaking the Western powers in industrial output, and
a corresponding leap in mathematics education was prescribed. Indeed, for years the Chinese Communist state
drilled into its people the urgency of raising their
country to a world military and industrial power. With
the advent of the Cultural Revolution, this emphasis
changed. Now the paramount concern of China's leaders
is the preservation and promotion of a Marxist-Leninist,
proletarian society. Chinese education has once more
been turned inward and the concept of mathematics as a
computational scheme designed to satisfy social needs
resurrected. In this respect, Chinese educational
reforms in mathematics have fitted successfully within

the plans for national development. The limited information available concerning the present content and pedagogy of mathematics teaching in the People's Republic of China indicates success in the proletarianization of mathematics teaching. Graduates of this system will possess a functioning knowledge of basic mathematics principles, while their ability to generalize mathematics concepts and undertake higher mathematical studies remains questionable. The fact that under this policy so many Chinese are becoming mathematically literate is worth noting. Perhaps the quality of the mathematical knowledge presently being taught is debatable, but the Chinese are not concerned, for the time being, with quality. The accomplishments of candidates on the Mathematical Olympiads have shown that the Chinese students can achieve a high level of mathematical proficiency. What is important at this time is that as much of the Chinese society as possible achieves some appreciation of the power of mathematics so that this generation of worker-peasants, having been thoroughly exposed to the broad aspects of mathematics, will bring forth a generation intent on expanding this knowledge. Then a whole society, rather than a select few, will advance in mathematical potential, and through it the industrial potential of the People's Republic of China will become most formidable!

BIBLIOGRAPHY

Arens, Richard. "The Impact of Communism on Education in China, 1949-50." Unpublished Ph.D. dissertation, University of Chicago, 1952.

Arndt, C. O. et al. Education in China Today. Washington, D. C.: U. S. Office of Education, 1944.

Barendsen, Robert. "Education in China: A Survey," Problems of Communism (1964), 13: 19-27.

Barendsen, Robert. "Half-Work, Half-Study Schools in Communist China," O. E. 14100. Washington, D. C.: U. S. Department of Health, Education and Welfare, 1964.

Barendsen, Robert. "Planned Reforms in the Primary and Secondary School System in Communist China," Education Around the World. Washington, D. C.: U. S. Department of Health, Education and Welfare, (1960), p. 6.

Becker, C. H. The Reorganization of Education in China. Paris: League of Nations, Institute of Intellectual Cooperation, 1932.

Biggerstaff, Knight. "The T'ung Wen Kuan," The Chinese Social and Political Review (1934), 18: 239.

Biggerstaff, Knight. The Earliest Modern Government Schools in China. Ithaca, New York: Cornell University Press, 1961.

Bo, C. S. Programs of New Education. Shanghai: People's Publishing House, 1949.

Bodde, Derk. Peking Dairy 1948-49: A Year of Revolution. New York: Fawcett Publishing Company, 1967.

Boyer, Carl B. A History of Mathematics. New York: John Wiley and Sons, 1968.

Breslich, Ernst R. and Stone, Charles. Trigonometry and Tables. New York: Ginn and Company, 1915.

Breslich, Ernst R. and Stone, Charles. Trigonometry with Tables for Use in Senior High Schools and Junior Colleges. Chicago: University of Chicago Press, 1928.

Bridgeman, E. C. Chinese Chrestomathy. Macao: Mission Press, 1841.

Cameron, Meribeth E. The Reform Movement in China. New York: Octagon Books, Inc., 1963.

Camman, Schuyler. "The Evolution of Magic Squares in China," American Oriental Society Journal (1960), 80: 116-24.

Canton Christian College, Bulletin 26. Canton: 1920.

"Chairman Mao on Revolution in Education," translated in Current Background. Hong Kong: American Consulate (August 22, 1969).

Chang, Chien. "Schooling for the Millions," China Reconstructs (October, 1959), p. 56.

Chang, Jen-chi. Pre-Communist China's Rural School and Community. Boston: The Christopher Publishing House, 1960.

Chang, Peng Chun. Education for Modernization in China. New York: Teachers College Press, 1923.

Ch'ang-shou, Spare-time Normal School, Chungking Mathematics Faculty Research Section. "Arrangement of Text Material on Mathematics in Spare-time Normal Schools," Shuxue Tongbao (July, 1965), pp. 2-3.

Chao Ts'u-keng. "On the Problems Adopted for the 1963 Mathematics Contest for Peking Middle School Students," Shuxue Tongbao (July, 1963), pp. 8-14.

Chapman, T. W. Middle School Arithmetic. Shanghai: Chung Hwa Book Company, 1919.

Chen, Anthony. "The Philosophy of Communist China as Applied to Secondary and Higher Education." Unpublished Master's thesis, De Paul University, Chicago, 1956.

Chen, An Ren. Modern Political History. Shanghai: Chung Hwa Book Company, 1943.

Ch'en, Ch'uan-kui, et al. "Experimental Teaching and Results of 'Second Degree Equations' of Algebra in Fifth Year Primary School Class," Shuxue Tongbao (June, 1960), pp. 7-11.

Chen, Hsuan-shan. "Training Teachers for Middle School," China Reconstructs (September, 1956), p. 19.

Chen, Pai-lin and Mao, Yu-yen. "On the Processes of Mastery of Typical Verbal Arithmetical Problems in School Children," Hsin-li Hsüeh-pao (No. 3, 1965), pp. 215-222.

Ch'en, Ta-jou. "Annual Academic Meeting of the Chinese Psychological Society," Hsin-li Hsüeh-pao (No. 1, 1964), pp. 109-112.

Chen, Theodore. "Education and Propaganda in Communist China," The Annals of the American Academy (September, 1951), 227: 135-145.

Chen, Theodore. "Educational Crisis in China," Educational Administration and Supervision (December, 1949), 34: 468-478.

Chen, Theodore H. E. "New Citizens for a New Society," Current History (September, 1965), 49: 155-163.

Chen, Theodore. Teacher Training in Communist China. Studies in Comparative Education Series, No. OE 14058. Washington, D. C.: U. S. Office of Education, 1960.

Chen, Theodore. "The New Education in Communist China," School and Society (March 18, 1950), 71: 166-169.

Cheng, J. C. "Notes and Comment: Half-Work and Half-Study in Communist China," Pacific Affairs (June, 1959), 32: 178-193.

Cheng, Tsu-hsin. "An Experimental Study of the Effect of Children's [Mental] Activity on Solving Verbal Problems in Division," Hsin-lin Hsüeh-pao (No. 4), 1964, pp. 369-376.

Chi, Tung-wei. Education for the Proletariat in Communist China. Hong Kong: Union Research Institute, 1956.

Chien, Kao. "Progressive Education Undermined China," The Freeman (December, 1954), pp. 216-218.

Chin, Robert and Chin, Ai-li S. Psychological Research in Communist China: 1949-1966. Cambridge, Mass.: M.I.T. Press, 1969.

China Handbook 1952-53. Taipei: China Publishing Company, 1954.

China News Agency. "First Batch of Students Graduated from Shanghai Television University," Surveys of China Mainland Press, No. 36313 (February 5, 1966), p. 5.

China News Agency. "Middle School Teaching Material Compiled Under Working Class Leadership," Shenyang (March 18, 1969); in Surveys of China Mainland Press, No. 4383 (March 25, 1969).

China News Agency. "Peking Middle School Pushes Forward Educational Revolution," Peking (May 15, 1970); Surveys of China Mainland Press, No. 4663 (May 25, 1970), p. 16.

China News Agency. "Poor and Lower-Middle [Class] Peasants Compile Teaching Material," Peking (January 17, 1969); in Surveys of China Mainland Press, No. 4343 (January 22, 1969), pp. 24-26.

China News Agency. "Revolution in Education Brings About New Outlook," Peking (February 2, 1969); in Surveys of China Mainland Press, No. 4355, p. 17.

China News Analysis. No. 81 (April 29, 1955).

Chin-jih Hsin-wen. May 7, 1966 in S.C.M.P 36409.

Ch'iu Hsueh-hua. "A Study of the Skill of Calculation with the Abacus in School Children," Hsin-lin Hsüeh-pao (No. 2, 1963), pp. 107-112.

Chou, Hsueh-Chli. "Some Basic Appreciation as to Developing Student's Ability to Relate Theory to Practice in Mathematical Thinking," Shuxue Tongbao (January, 1966), pp. 3-7.

"Chronology of the Two-Road Struggle on the Educational Front in the Past Seventeen Years," translated in Chinese Education (Spring, 1968), pp. 4-50.

Clyde, Paul and Beers, Burton. The Far East. Englewood Cliffs, New Jersey: Prentice Hall, 1966.

Compilation of Problems from 1956-57 Mathematics Competitions for Middle School Students in Shanghai Municipality. Shanghai: New Knowledge Press, 1958.

"Computer Developments," Shuxye Tungxyn (December, 1957).

"Conclusions of the 1962 Mathematics Contest Among Middle
    School Students in Peking Municipality," Shuxue
    Tongbao (Apeil, 1963), pp. 50-51.

Corbett, Charles H. Shantung Christian University. New
    York: United Board for Christian Colleges in China,
    1955.

Darrock, John. "Chinese Textbooks," Journal of North China
    Branch of the Royal Asiatic Society (1906), 37: 208-
    214.

DeFrancis, John. "Mathematical Competitions in Communist
    China," The American Mathematical Monthly (1962), 55:
    251-255.

DeFrancis, John. "The Mathematics Scene in China," The
    Mathematics Teacher (April, 1962), 55: 251-255.

De Morgan, August. Elements of Algebra Preliminary to the
    Differential Calculus. London: Taylor and Watson,
    1837.

Dewey, John. The Dalton Laboratory Plan. New York:
    Dutton, 1921.

Dewey, John and Dewey, Evelyn. Schools of Tomorrow. New
    York: E. P. Dutton and Company, 1915.

Doolin, Dennis and Ridley, Charles. The Genesis of a
    Model Citizen in Communist China. Washington, D. C.:
    U. S. Office of Health, Education and Welfare, 1968.

"Draft Program for Primary and Middle Schools in Chinese
    Countryside," Peking (May 13, 1969); in Surveys of
    China Mainland Press, No. 4418 (May 19, 1969), pp. 9-15.

Educational Association of China. The Outline Standards
    of the New System Curriculum. Shanghai: Commercial
    Press, 1925.

Fairbank, John. China, The People's Middle Kingdom and
    the U.S.A. Cambridge, Mass.: Belknap Press of
    Harvard University, 1967.

Fairbank, John and Teng, Ssu. China's Response to the
    West: A Documentary Survey. Cambridge, Mass.:
    Harvard University Press, 1954.

Fairfax-Cholmeley, Elsie. "Education in China," New
    World Review (August, 1956), p. 37.

Fan, Chih-lung. "College Courses by Television," China
    Reconstructs (April, 1961), pp. 10-11.

Farm Village Practical Handbook. Shanghai: Shanghai
    Publishing Company, 1966.

First Middle School Research Group, Canton Province,
    Ching-min City. "Follow Party Rules to Revise Mathe-
    matics," Chung Hsueh Shu Hsueh (February, 1959), pp.
    19-20.

First Part-Work Part-Study Technical School of the First
    Bureau of Machine-Building Industry of Tientsin
    Municipality. "Mathematics Curricular Program for
    Part-Work Part-Study Middle Technical Schools,"
    Shuxue Tongbao (September, 1965), pp. 2-6.

Fisher, O. "Education in Communist China," School and
    Society (June 20, 1959), pp. 302-305.

Fraser, Stewart. Chinese Communist Education. Nashville:
    Vanderbilt University Press, 1963.

Fremantle, Anne. Mao Tse-tung; An Anthology of His
    Writings. New York: Mentor, 1962.

Fresson, Chester J. "The Course of Study in the Mission
    School," Educational Review, Shanghai (1909), 2: 1-12.

Fryer, John. "Science in China," Nature (May 19, 1881),
    p. 55.

Fu, Tsu-ch'ung. "Middle School Extracurricular Mathe-
    matics Activities," Shuxue Tongbao (April, 1963),
    pp. 2-4.

Fuh Tan University. Fuh Tan University Catalogue and
    Directory 1909-11. 1909.

Fuh Tan University. Fuh Tan University Catalogue and
    Directory 1917-18. 1917.

Galt, Howard. The Development of Chinese Educational
    Theory. Shanghai: The Commercial Press, 1929.

Goodrich, L. Carrington. A Short History of the Chinese
    People. New York: Harper and Row, Publishers, 1950.

Granville, William A. Plane Trigonometry and Four Place
    Tables of Logarithms. New York: Ginn and Company,
    1909.

Gressy, Earl H. and Chih, C. C. East China Studies in
    Education, No. 5: Middle School Standards. Shanghai:
    East China Christian Association, 1929.

Halmos, P. R. "The Meaning of the Spectrum Theorem," Shuxue Tongbao (October, 1964), pp. 47-49.

Han, Erh-tsai. "They Like Mathematics," China Reconstructs (December, 1962), 11: 34-35.

Harner, Evelyn. Middle School Education as a Tool of Power in Communist China. Santa Barbara, Calif.: General Electric Company, 1962.

Hermann, H. "On the Surveying Instruments Described in Parker's Trigonometry," Educational Review (August, 1910), 3: 1-3.

Herschel, John. Outlines of Astronomy. London: Longman, Green and Roberts, 1859.

Hewitt, Edwin. "Application of Connectivity in Analysis," Shuxue Tongbao (March, 1965).

Ho, Tso-ch'uang. "The Function of the Mathematical Method in the Cognition of the Objective World," Hung-ch'i (May 16, 1962.

Hobson, William E. A Treatise on Plane Trigonometry. Cambridge: Cambridge University Press, 1897.

"How a Computer Works," Shuxye Tungxyn (June, 1958), pp. 6-8.

Hsiao, Chien-ying. "The Characteristics of the Thinking Process of Solving Verbal Arithmetical Problems by First Grade Children," Hsin-li Hsueh-pao (No. 1, 1965), pp. 50-56.

Hsu, Chien-hua. "My Views on the Surveying Practice for Middle School Students in Studying Mathematics," Shuxue Tongbao (November, 1965), pp. 20-21.

Hu, Chang-tu. "Recent Trends in Chinese Education," International Review of Education (January, 1964), 10: 12-21.

Hu, Chang-tu. Chinese Education Under Communism. New York: Teachers College Press, 1962.

Hu, Yen-li. How to Carry Out Five Loves Education. Peking, 1951.

Hua, Lo-kĕng. Additive Prime Number Theory. Peking: Chinese Academy of Sciences, 1953.

Hua, Lo-kêng. "Chairman Mao Points Out the Road of
    Advance for Me," China Reconstructs (November, 1969),
    pp. 30-31 and 41.

Hua, Lo-kêng. "Completion of the Peking Competition,"
    Shuxue Tongbao (June, 1956), pp. 1-2.

Hua, Lo-kêng. Harmonic Analysis of Functions of Several
    Complex Variables in Classical Domains. Moscow:
    Izdat. Inostr. Lit., 1959.

Hua, Lo-kêng. To a Young Mathematician. Shanghai:
    China Youth Press, 1956.

Hua, Lo-kêng. "We Will Have National Mathematics Com-
    petitions Soon," Shuxue Tongbao (January, 1956), pp.
    1-3.

Hua, Lo-kêng and Wang, Yuan. Classical Groups. Shanghai:
    Shanghai Science and Technology Press, 1963.

Hua, Lo-kêng and Wang, Yuan. "On the Problem of Calculat-
    ing Mineral Reserves and Hill Area on Contour Line
    Maps," Acta Mathematica Sinica, 1961, Vol. 1.

Hua, Lo-kêng, et al. "Applications of Mathematical
    Methods to Wheat Harvesting," Acta Mathematica Sinica,
    1961, 1: 77-91.

Huang, Kuo-pao. "The Second Congress of the Chinese
    Mathematics Society," Shuxue Tongbao (April, 1960),
    pp. 2-4.

Huang, Sung-men. "On After-Hour Mathematics Study
    Groups," Shuxue Tongbao (November, 1964), pp. 17-22.

Hughes, E. R. The Invasion of China by the Western
    World. London: Adam and Charles Black, 1938.

Hung, Guang-shun. "Extracurricula Mathematics," Shuxue
    Tongbao (May, 1956), pp. 39-40.

Hunter, Edward. Brainwashing in Red China. New York:
    Vanguard Press, 1951.

Huo, Mao-cheng, et al. "Teaching Algebra in Primary
    Schools Experimentally," Shuxue Tongbao (April, 1960),
    pp. 14-18.

Hwa, Yu. "China," Christian Century (September 30, 1970), p. 1170.

Jen-min Jih-pao. "Aomen Road No. 2 Primary School, Shanghai Puts Extracurricular Activities on the Agenda of its Revoluntary Committee" (July 10, 1970); in Surveys of China Mainland Press, No. 4706.

Jen-min Jih-pao. "Grasp Extramural Education" (August 27, 1969), pp. 1-3.

Jen-min Jih-pao. "Hung Ch'i Middle School in Penhsi, Liaoning, Forms Textbook Group (May 8, 1969); translated in Surveys of China Mainland Press, No. 4419 (May 20, 1969), pp. 6-7.

Jen-min Jih-pao. "Several Arithmetical Questions" (January 14, 1969); translated in Surveys of China Mainland Press, No. 4355 (February 7, 1969), p. 12.

Jen-min Jih-pao. "Shihchingshan Middle School Establishes the Idea of 'Study for the Revolutionary Cause'" (May 15, 1969); translated in Surveys of China Mainland Press, No. 4423 (May 26, 1969), pp. 6-7.

Kiang, Ying Cheng. "The Geography of Higher Education in China." Unpublished Ph.D. dissertation, Columbia University, New York, 1955.

Kikuchi, Baron. Japanese Education. London: John Murray, 1909.

King, Edmund J. Communist Education. New York: Bobbs Merrill Company, 1963.

King, H. E. The Educational System of China as Recently Constructed. Bulletin 1911, No. 15, Washington, D. C.: Government Printing Office, 1911.

Kolatsunowa, V. G. "The Procession of a Mathematics Evening Party," Shuxue Tongbao (October, 1962), pp. 7-8.

Ku, Ch'ing. "The Problems Existing Among Students as Found From a Test and How to Improve the Quality of Mathematics Teaching," Shuxue Tongbao (March, 1969), pp. 12-15.

Kuang, Han. "On Whether it is Necessary to Re-write the Mathematics Outlines for Elementary and Middle Schools," Shuxue Tongbao (January, 1960), pp. 32-35.

Kuang-ming Jih-pao. "Kirin Normal University Reforms Mathematics and Education Systems," Peking (April 27, 1960), p. 2; translated in Joint Publications Research Service, No. 15, 515.

Kuang-ming Jih-pao. "Persevere in and Intensify the Educational Revolution," Peking (March 20, 1970); translated in Surveys of China Mainland Press, No. 4631 (April 8, 1970), pp. 60-71.

Kuang-ming Jih-pao. "Students Assisted in Their Creative Study and Creative Application of Chairman Mao's Works Outside the School," Peking (January 14, 1969), pp. 3-4; S.C.M.P. 4351.

Kuang-ming Jih-pao. March 8, 1961.

Kuang-ming Jih-pao. November 11, 1961, in S.C.M.P. 2654.

Kuang-ming Jih-pao. March 9, 1962.

Kuang-ming Jih-pao. March 12, 1962, in Surveys of China Mainland Press, No. 2710. Hong Kong: United States Consulate General, April 2, 1962.

Kuo, Ping Wen. The Chinese System of Public Education. New York: Teachers College Press, 1914.

Lee, Yee-chung. "Expressing My Opinion on Party Policy [in Mathematics Education]," Chung-hsüeh Shu-hsüeh (January, 1959), pp. 2-4.

Li, Hsi-yen. "On Some Problems Associated with the Teaching of Polar Coordinates in Middle School Geometry," Shuxue Tongbao (June, 1964), pp. 15-20.

Liehmulu, V. G. "Geometric Explanation of Abnormal Phenomena in Rotation of Hyperbola and Special Theory of Relativity," Shuxue Tongbao (March, 1965), pp. 39-42.

Lin, Ch'eng-chu. "My Views on Middle School Pedagogical Reform," Shuxue Tongbao (May, 1960), pp. 17 and 41.

Lindsay, Michael. Notes on Educational Problems in Communist China 1941-47. New York: Institute of Pacific Relations, 1950.

Liu, Ching-ho. "A Preliminary Discussion on Some of the Problems Regarding the Learning Process of Children," Jen-min Jih-pao (March 26, 1961); translated in Joint Publications Research Service, No. 9398.

Loomis, Elias. Elements of Differential and Integral Calculus. New York: Harper and Brothers, 1874.

Loomis, Elias. Elements of Geometry and Conic Sections. New York: Harper and Brothers, 1847.

Lu, Ching and Wang, Wen-chun. "The Thinking Processes in Arithmetical Operations in Primary School Children," Hsin-li Hsüeh-pao (No. 2, 1960), pp. 121-135.

Lu, Ching et al. "The Mastery of Reverse Operation and Flexibility of Thinking in Arithmetic in School Children," Hsin-li Hsüeh-pao (No. 3, 1963), pp. 237-247.

Lu, Chung-heng and Chu, Hsin-ming. "The Effect of Outer Shape on the Perceptive Structure of Geometric Figures and Thought Processes," Hsin-li Hsüeh-pao (No. 3, 1964), pp. 248-257.

Lu, Chung-heng et al. "Some Psychological Factors in Promoting the Student's Grasp of Arithmetical Knowledge, as found in Recent Educational Reforms," Hsin-li Hsüeh-pao (No. 3, 1961), pp. 190-201.

Lu, Ting-yi. "Education Must be Combined with Productive Labor," in Stewart Fraser, Chinese Communist Education. Nashville: Vanderbilt University Press, 1963, pp. 283-300.

Lu, Ting-yi. "Education Must be Reformed," translated in Current Background, No. 630. Hong Kong: United States Consulate General (August, 1960).

Mac Farquhar, Roderick. The Hundred Flowers Campaign and the Chinese Intellectuals. New York: Fredrick Praeger, Inc., 1960.

Mao Tsê-tung. "Comment on 'Views Advanced by a Middle School Principal Concerning the Question of Alleviating the Work Load of Middle School Students'," translated in Current Background (March 10, 1964).

Mao Tsê-tung. "Instruction of the Military Affairs Committee on the Question of Consolidating the Anti-Japanese Military and Political College," translated in Current Background. Hong Kong: American Consulate General (August 22, 1969), p. 4.

Mao Tsê-tung. "Method of Work Draft," translated in Current Background, No. 888. Hong Kong: United States Consulate General (August 22, 1969).

Mao Tsê-tung. "Resolution of the 9th Congress of the 4th Army of the Red Army of the Communist Party of China," translated in Current Background. Hong Kong: American Consulate General (August 22, 1969), p. 2.

320

Mao Tsê-tung. "The United Front in Cultural Work,"
    translated in Current Background. Hong Kong: Ameri-
    can Consulate General (August 22, 1969), p. 6.

Mao, Yu-yen, et al. "The Relation Between the Structure,
    Formulation of the Arithmetical Problems and the
    Psychological Activity of the Pupils in the Process
    of Problem Solving," Hsin-li Hsüeh-pao (No. 4, 1965),
    pp. 291-297.

Martin, W. A. P. A Cycle of Cathay. London: Olephant,
    Anderson and Ferrier, 1900.

Martin, W. A. P. Hanlin Papers. Shanghai: Kelly and
    Walsh, 1880.

"Mathematics Outline for Part-time Agricultural Middle
    School," Shuxue Tongbao (February, 1966), pp. 9-10.

Mathematics Teaching and Research Team, Tsinchow Textile
    and Machine Manufacturing School. "Spare-time
    Mathematics Outline for Tsinchow Textile and Machine
    Manufacturing School," Shuxue Tongbao (January, 1959),
    pp. 26-29.

Mathematics Section, Second Middle School, Canton. "Com-
    bine Mathematics Teaching with the Present Situation,"
    Chung-hsüeh Shu-hsüeh, March, 1959, pp. 1-2.

Mikami, Yoshio. The Development of Mathematics in China
    and Japan. New York: Chelsea Publishing Company,
    reprint of 1913 edition.

Ministry of Education, People's Republic of China.
    "Introduction of Teachers College Mathematics Train-
    ing Program," Chung-hsüeh Shu-hsüeh (November, 1956),
    pp. 41-45.

Ministry of Education, People's Republic of China.
    "Mathematics Schedule for the Junior Normal College,"
    Chung-hsüeh Shu-hsüeh (November, 1956), pp. 37-40.

Ministry of Education, People's Republic of China.
    "Middle School Mathematics Teaching Outline: 1956-
    57," Shuxue Tongbao (August, 1956), pp. 24-39.

Moehlman, A. H. and Rouceh, J. S. Comparative Education.
    New York: Dryden Press, 1952.

Morgan, L. G. The Teaching of Science to the Chinese.
    Hong Kong: Kelly and Walsh Ltd., 1933.

321

Morrison, Ester. "A Comparison of Kuomintang and Commu-
    nist Modern History Books," Papers on China, Harvard
    Seminars (March, 1952), pp. 3-45.

Myrdal, Jan. Report From a Chinese Village. New York:
    Pantheon Books, 1965.

Needham, Joseph. Science and Civilization in China.
    4 vols. London: Cambridge University Press, 1959.

New York Times. "China's New Math and Old Problems"
    (March 9, 1969), p. 18.

"1963 Peking Municipality Mathematical Competitions,"
    Shuxue Tongbao (May, 1963), back cover.

North China Herald. October 30, 1899.

Nunn, Raymond. Publishing in Mainland China. Cam-
    bridge, Mass.: M.I.T. Press, 1966.

"Office of Mathematics, Physics and Logic of the Insti-
    tute of Mathematics of the Chinese Academy of Science
    Sponsors Lectures on Mathematics Foundation," Shuxue
    Tongbao (February, 1960), p. 42.

Oldham, C. H. G. "Science and Education," Bulletin of
    Atomic Scientists (June, 1966), p. 43.

Orleans, Leo. Professional Manpower and Education in
    Communist China. Washington, D. C.: National
    Science Foundation, 1961.

"Outline of Examination in Mathematics for Matriculation
    to Institutions of Higher Education in the People's
    Republic of China for 1959," Current Background
    (August 17, 1959), pp. 6-12.

Pan, Shu. "Ideas About Expanded Research in Educational
    Psychology," Kuang-ming Jih-pao (March 13, 1962);
    translated in Joint Publications Research Service,
    No. 13531.

Peake, Cyrus. Nationalism and Education in Modern China.
    New York: Columbia University Press, 1932.

"Peking City 1957 College Entrance Results in Mathematics,"
    Shuxue Tongbao (May, 1958), pp. 12-13.

Peking Normal University, Mathematics Education Research
    Group. "Problems in Current Middle School Mathematics,"
    Shuxue Tongbao (March, 1960), pp. 30-32.

Peking Normal University, Mathematics Faculty Research Section, Second Middle School. "Conducting Active Extracurricular Activities," Shuxue Tongbao (February, 1963), pp. 2-4.

Peking Normal University, Middle and Primary School Mathematics Pedagogical Reform Research Group. "Suggestions on the Modernization of the Mathematics Curricular Materials for Middle and Primary Schools," Shuxue Tongbao (April, 1960), pp. 4-10.

Peking Normal University, Middle School Mathematics Teaching and Research Group. "Our Ideas Concerning Present Middle School Mathematics Teaching," Shuxue Tongbao (January, 1960), pp. 35-38.

Peking Normal University, Science Research Group, Freshman Class. "Our Views on the Problem of Dovetailing Mathematics Courses in Middle Schools with that in Colleges," Shuxue Tongbao (March, 1960), pp. 35-37.

Peking Normal University, Spare-time Middle School Mathematics Editing Section. "Spare-time Mathematics Teaching Outline," Shuxue Tongbao (October, 1958), p. 23.

Pien, Shu-yang, et al. "Opinions on the Goals and Mission of Middle School Mathematics," Shuxue Tongbao (January, 1960), pp. 38-40.

Price, R. F. Education in Communist China. New York: Praeger Publishers, 1970.

Priestly, K. E. Education in China. Hong Kong: Dragonfly Books, 1961.

Purcell, Victor. Problems of Chinese Education. London: Kegan, Paul and Trench, Trubner and Company, 1936.

"Put Mao Tse-tung Thought in Command of Cultural Courses," Red Flag, Peking (March, 1971).

"Reference Books Used in the Past Ten Years," Shuxue Tongbao (October, 1959), pp. 26-29.

"Required Subjects of Six Normal Schools in China," The Educational Magazine, Shanghai (May, 1922), pp. 1-43.

"Results of the Mathematics Contest for Middle School Students in Peking Municipality," Shuxue Tongbao (April, 1963), pp. 50-51.

Ross, Edward Alworth. The Changing Chinese. New York:
  The Century Company, 1911.

Selden, Mark. "Yenan Communism: Revolution in the
  Shensi-Kansu-Ninghsia Border Region 1927-1945."
  Unpublished Ph.D. dissertation, Yale University, 1967.

Seybolt, Peter. "Yenan Education and the Chinese Revolution." Unpublished Ph.D. dissertation, Harvard
  University, 1969.

Shanghai Normal University, Mathematics Department.
  "Four Year System Middle School Mathematics Teaching
  Outline," Shuxue Jeaoxue (April, 1959), pp. 6-11.

Shang-hsi shih, 1956-57 nien Chung hsüeh-sheng Shu-hsüeh
  Cheng-sai his-t'i pien-hui [Compilation of Problems
  from 1956-57 Mathematics Competitions for Middle
  School Students in Shanghai Municipality]. Shanghai:
  New Knowledge Press, 1958.

Shangtung Christian University Catalogue 1928, Bulletin
  No. 57. Shangtung: 1928.

Shantung Provincial Department of Education. "Demolish
  the 'Little Treasure Pagoda,' System of Revisionist
  Education," Jen-min Jih-pao (December 17, 1967).

Shensi-Kansu-Liangshung Border Area Education Department.
  Primary Arithmetic. 6 vols. Shen-fu Bookstore, 1946.

Shih, Lui. "China's New Educational System," People's
  China (December 1, 1951), p. 7.

Shklarsky, D. O., Chentzov, N. N., Yaglom, I. M. The
  U.S.S.R. Problem Book: Selected Problems and Theorems
  of Elementary Mathematics. San Francisco: W. H.
  Freeman and Company, 1961.

Shuxue Jiaoxue. April, 1959, pp. 27-28.

Shuxue Tongbao. April, 1956.

Shuxue Tongbao. January, 1960.

Shuxue Tongbao. September, 1962.

Simpson, R. F. "The Development of Education in Mainland
  China," Phi Delta Kappan (December, 1957), pp. 84-93.

Smith, David E. History of Mathematics. 2 vols. New
  York: Dover Publications, reprint of 1923 edition.

Smith, Harold F. Elementary Education in Shangtung China.
  New York: Columbia University Press, 1930.

Sokokusha, The. Anti-Foreign Teaching in New Textbooks of China. Tokyo: 1929.

Soochow University Catalogue 1934. Soochow: 1934.

Speiser, A. Theorie der Gruppen von endlicher Ordnung. Basel: Birkhauser, 1956.

St. John's University Catalogue 1923. Shanghai: Presbyterian Mission Press, 1923.

State Statistical Bureau, People's Republic of China. Ten Great Years. Peking: People's Press, 1959.

"Strive to Build a Socialist University of Science and Engineering," Hung Ch'i (August, 1970).

"Summer 1955 Senior Middle School Entrance Examination in Mathematics," Shuxue Tongbao (August, 1955), pp. 47-48.

Surveys of China Mainland Press, No. 2270. Hong Kong: United States Consulate General, 1960.

Swetz, Frank. "Training of Mathematics Teachers in the People's Republic of China," American Mathematical Monthly (December, 1970), pp. 1097-1103.

Tai, Shu. Self Teaching Algebra. 2 vols. Shanghai: Shanghai Arts and Science Press, 1964.

Tan, Jen-mai. "History of Modern Chinese Secondary Education." Unpublished Ed.D. dissertation, University of Pennsylvania, 1940.

Tan, W. T., et al. Education in China 1924. Shanghai: The Commercial Press, 1925.

Tawney, R. H. Land and Labour in China. London: George Allen and Unwin Ltd., 1932.

Teng, T. T. and Lew, T. T. Education in China. Peking: The Society for the Study of International Education, 1923.

Tengchow College Catalogue 1891. American Presbyterian Press [China], 1891.

Terman, E. L. The Efficiency of Elementary School in China. Shanghai: The Commercial Press, 1924.

Tsang, Chiu-sam. Society, Schools and Progress in China. New York: Pergamon Press, 1968.

Tsi, Pei-ting and Chan, Tse-min. "Middle School Survey Practice," Shuxue Tongbao (May, 1957), pp. 32-37.

Tuan, Hsueh-fu. "Learn from Russia to have Mathematical Competitions," Shuxue Tongbao (January, 1956), pp. 3-5.

Tuen, Lo King and Graybill, H. B. Practical Arithmetic: A Mastery of Modern Mathematics for Chinese Middle Schools, Commercial Schools and Business Students. Shanghai: Edward Evans and Sons, 1923.

Twiss, George R. Science and Education in China. Shanghai: Commercial Press, 1925.

Union Research Service. Union Research Service Notes April-June 1960. Vol. 19. Hong Kong: Union Research Service, 1960.

United Nations. Economic and Social Council. World Survey of Education. 2 vols. Paris: 1958.

United States Department of State. United States Consular Reports. Washington, D. C.: January, 1902.

University of Nanking Catalogue 1931. Nanking: 1931.

"Using Mao Tse-tung's Thought to Occupy the Positions Outside the School," Surveys of China Mainland Press, No. 4673 (June 10, 1970).

Vanhee, Louis. "The Great Treasure House of Chinese and European Mathematics," The American Mathematical Monthly (1926), 33: 502-506.

Wang, Chih-heng. "Our Mathematical Land," Shuxue Tongbao (September, 1964).

Wang, Hsien-tien, et al. "An Investigation into the Development of Concepts in Children 4-9 Years [Old]," Hsin-li Hsüeh-pao (No. 4, 1964), pp. 352-360.

Wang, Li-keng. Arithmetic Stories. Peking: Shing-Hwa Book Store, 1958.

Wang, Ling and Needham, Joseph. "Horner's Method in Chinese Mathematics: Its Origins in the Root Extraction Procedures of the Han Dynasty," T'oung Pao (1955), 43: 343-388.

Wang, Y. C. Chinese Intellectuals and the West, 1872-1947. Chapel Hill: University of North Carolina Press, 1966.

Wang, Yu-ch'ing. "How to Teach the First and Second Chapters of Junior School Algebra," Shuxue Tongbao (February, 1960), pp. 17-21.

Wei, Chang, et al. "The Development of the Ability to Compare in Primary School Children," Hsin-li Hsüeh-pao (No. 3, 1964), pp. 274-280.

Webster, Norman. "China's Factories Mind the Children," The New York Times (November 19, 1970), p. 56.

"West China Union Course of Study," Educational Review. Shanghai (1911), 4: 12-21.

Weyl, H. Symmetry. Princeton: Princeton University Press, 1952.

Whewell, William. The Mechanical Euclid. Cambridge: Cambridge University Press, 1843.

Wilson, Dick. Anatomy of China. New York: Weybright and Talley, Inc., 1966.

Wu, Ch'i-ch'i. "The Contents of Mathematics Evening Parties," Shuxue Tongbao (October, 1962), pp. 6-7.

Wu-Gong. "To Realize Educational Policy from Extracurricular Activities," Zhongxue Shuxue (September, 1959), pp. 7-8.

Wu, Hsueh-lu. "My Experience in Improving Mathematics Pedagogy According to Chairman Mao's Ideology," Shuxue Tongbao (February, 1966), pp. 4-6.

Yan, Phou-lee. When I Was a Boy in China. Boston: Lothrop Publishing Company, 1887.

Yang, Thaddeus. "The Development of Education in China." Unpublished Master's thesis, De Paul University, 1951.

Yang-ts'un Normal School, Hopeh. "Preliminary Views on Mathematics Pedagogy in Four Year Part-Time Normal Schools," Shuxue Tongbao (August, 1965), pp. 2-6 and 20.

Yang, Yu-men, et al. "Experimental Teaching and Results 'Limits' of Algebra in Second Year Junior Middle School Class," Shuxue Tongbao (June, 1960), pp. 11-13.

Yang, Yung-hsiang. Self Teaching Plane Geometry. 2 vols. Shanghai: Shanghai Arts and Science Press, 1964.

Yao, Chien-ch'u and Yu, Yi-tze. Self Teaching Trigonometry. Shanghai: Shanghai Arts and Science Press, 1963.

Yau, C. S. Analysis of Correspondence Teaching of Mathematics in the Soviet Union for Forty Years. Shanghai: Science Publishing Press, 1965.

Yeh, Wei. Slave Education of the Middle School in Communist China. Hong Kong: Freedom Press, 1951.

Yen, Sun-ho. Chinese Education from the Western Viewpoint. New York: Rand McNally Company, 1913.

Ying, Yu-yeh, et al. "The Negative Effects of Crossing of Geometrical Figures on the Perceptual and Thought Processes," Hsin-li Hsüeh-pao (No. 1, 1964), pp. 72-93.

Yu, Chun-hsien. A Method of Measurement. Shanghai: Commercial Press, 1952.

Yu, Lan. "Why China Has No Science--An Interpretation of the History and Consequences of Chinese Philosophy," The International Journal of Ethics (1922), 32: 237-263.

Yuan, Pae-yeo. "A Comparative study of Chinese and American Secondary Education." Unpublished Master's thesis, University of Pennsylvania, 1924.

Zen, Wei-ts. "The Role of Education in Postwar China." Unpublished Ed.D. thesis, University of Pennsylvania, 1948.

Zi, Le P. Etinne. Pratique des Examens Litteraires en Chine. Shanghai: Catholic Press, 1884.

APPENDICES

APPENDIX A

AMERICAN MATHEMATICS TEXTS POPULAR IN CHINA
IN THE YEARS 1925-1949

Subject	Title	Author(s)
Algebra	Algebra*	Wentworth
	College Algebra	Davisson
	Higher Algebra	Knight and Hall
	Algebra	Hawkes, Luby, Touton
	School Algebra	Wentworth
	Second Course in Algebra	Wills and Hart
	Academic Algebra	Wentworth and Smith
	First Course in Algebra	Ford and Ammerman
	High School Algebra	Milne
	Elementary Algebra	Baker and Bourne
Arithmetic	Primary Arithmetic	Wentworth
	Grammar School Arithmetic	Wentworth
	Complete Arithmetic	Wentworth and Smith
	High School Arithmetic	Wentworth and Hill
	Essentials of Arithmetic	Bergstresser
Geometry	Plane and Solid Geometry*	Wentworth
	Solid Geometry*	Wentworth
	Plane and Spherical Geometry	Bauer and Brooke
	Plane Geometry	Young and Jackson
	Elements of Geometry	Phillips and Fisher
	Plane Geometry	Stone and Millis
	Plane Geometry	Berman and Smith
	Plane Geometry	Wentworth and Smith
	Practical Exercises in Geometry	Eggar
	Plane and Solid Geometry	Berman and Smith
Integrate Mathematics	Mathematics (Book I and II)*	Breslich
	General Mathematics	Shorling and Reeve
	Junior High School Mathematics	Wentworth, Smith, Brown

APPENDIX A (Continued)

Subject	Title	Author(s)
	Mathematics for Secondary Schools	Mayers et al.
Trigonometry	Plane and Spherical Trigonometry	Kenyon and Ingold
Other	Analytic Geometry	Love
	Elementary Functions and Their Applications	Gale and Watkey
	Introduction to Elementary Functions	McClenon
	Business Arithmetic	More and Miner
	Concise Business Arithmetic	More and Miner

* Denotes books translated into Chinese.

APPENDIX B

A PARTIAL LISTING OF HIGHER NORMAL INSTITUTIONS
FOR THE TRAINING OF MATHEMATICS TEACHERS
IN THE PEOPLE'S REPUBLIC OF CHINA*

Province or Region	Institution
Anhwei	Anhwei Normal College, Wuhu Hofei Special Normal School, Hofei
Chekiang	Chekiang Normal College, Hangchow Special Normal School, Hangchow
Fukien	Fukien Normal College, Foochow
Heilungkiang	Harbin Normal College, Harbin Pedagogical Radio University, Harbin (founded 1960, broadcasts mornings and evenings; by 1962 it had 6000 students)
Honan	Second Honan Normal College, Hsinhsiang First Honan Normal College, Kaifeng, Kaifeng Special Normal School, Kaifeng
Hopei	Hopei Normal College, Paoting Shihchiachuang Normal University, Shihchiachuang Tientsin Pedagogical College Tientsin Peking Normal University, Peking Peking Normal College, Peking
Hunan	Hunan Normal College, Changsha Changsha Special Normal School, Changsha
Hupeh	Central China Normal College, Wuhan Hupeh Special Normal School, Wuchang

APPENDIX B (Continued)

Province or Region	Institution
Inner Mongolia Autonomous Region	Inner Mongolia Normal College, Huhehot
Kansu	Northwestern Normal College, Lanchow
Kiangsi	Kiangsi Normal College, Nanchang Nanchang Special Normal School, Nanchang
Kaingsu	Kiangsu Normal College, Soochow Nanking Normal College, Nanking North Kiangsu Normal College, Yangchow East China Normal University Shanghai (correspondence program began in 1956, by 1962, 900 correspondence graduates) East China Normal College, Shanghai Shanghai Normal College No. 1, Shanghai Shanghai Normal College No. 2, Shanghai Shanghai Pedagogical University, Shanghai
Kirin	Kirin Normal University, Changchun (correspondence program began in 1953; 1960 instituted correspondence studies for teachers; by 1961 graduated 500 correspondence students) Northeast Normal University, Changshun Kirin Special Normal School, Changchun
Kwangtung	South China Normal College, Canton Canton Normal College, Canton Canton Special Normal School, Kuangchou
Kweichow	Kweiyang Normal College, Kweiyang

APPENDIX B (Continued)

Province or Region	Institution
Kwangsi-Chuang Autonomous Region	Kwangsi Normal College, Kweilin
Liasoning	Darien Normal College, Darien Darien Special Normal School, Darien Mukden Normal College, Mukden
Shansi	Shansi Normal College, Tasiyuan
Shantung	Shantung Normal College, Tsinan Chufou Normal College, Chufou
Shensi	Shensi Normal College, Sian Sian Normal College, Sian Normal College of Yenpien University, Yenchi
Sinkiang-Uighur Autonomous Region	Urumchi Normal College, Urumchi
Szechwan	Southwestern Normal College, Chungking Chungking Normal College, Chungking Chengtu Normal College, Chengtu Szechwan Normal College, Nanchung
Tibet	Tibet Normal College, Lhassa
Yunnan	Kunming Normal College, Kunming

* Compiled from: Leo Orleans, Professional Manpower and Education in Communist China (Washington, D. C.: National Science Foundation, 1961); "Teacher Training Institutes," pp. 194-201; Hans-Jurgen Eitner, Erziehung und Wissenschaft in der Volksrepublic China (Frankfurt: 1964), pp. 58-61; Chi-Wang, Mainland China's Organizations of Higher Learning in Science and Technology, Library of Congress Report (Washington, D. C.: U. S. Government Printing Office, 1961.

APPENDIX C

MIDDLE SCHOOL MATHEMATICS TEACHING SYLLABUS
1956-57*

Seventh Grade

Arithmetic   6 hr/wk   total 204 hrs.

1. Work with whole numbers   33 hrs.
   Review reading and writing of numbers; rounding
   off to the nearest 10, 100, 1000 etc.; metric
   system; solution of general problems involving
   the four operations; use of abacus, arithmetic
   mean; arithmetic principles and properties; use
   of parenthesis; measurement of time and time
   related problems; mental calculation; formation
   of simple equations; writing of simple budgets;
   areas and perimeters of square and rectangle;
   volume of cube and rectangular prisms.

2. Preparation for fractions   20 hrs.
   Fractions and their multiples; tests for division by 2, 3, 4, 5, 9, 25; concept of factoring; methods of finding: C.F., G.C.F., C.M.
   and L.C.M.

3. Fractions   67 hrs.
   Conversion of an improper fraction to a mixed
   number; ordering of fractions; general properties of fractions; reduction; four operations
   with fractions; ratio; proportion; percent;
   solution of word problems; scale drawing;
   surface area of cubes and rectangular prisms;
   area of triangle; construction of simple solid
   models.

4. Decimals   33 hrs.
   Definition of decimals; reading and writing;
   ordering; four operations with pencil and
   paper abacus and tables; solution of general
   problems; conversion of fractions to decimals;
   visual estimation of volume and distance confirmed by measurement; area and circumference
   of circles; volume and surface area of cylinder.

5. Percentage   12 hrs.
   Percentage as a relation between two numbers;
   use of percentage tables; solution of complex
   problems; construction of bar, line and pie
   graphs.

Seventh Grade

    6.   Ratio and inverse ratio   27 hrs.
          Properties of ratios; various ways of expressing ratios.

    7.   Survey practice   6 hrs.
          Construction of a line; measurement employing tape, pacing and compass; visual estimation of distance; use of layout square; layout of right angles; construction and area measurement of a quadrilateral.

    8.   Review   6 hrs.

Eighth Grade

    Algebra   4 hr/wk   total 136 hrs.

    1.   Equations and functions   20 hrs.
          Introduction to the use of variables in mathematics; application of all known mathematical operations to variable statements; tabulation of data tables; derivation of arithmetic equations; relation between dependent and independent variables; concept of function; solution of problems involving first degree equations.

    2.   Rational numbers   24 hrs.
          Development of integers; properties of zero, use of coordinates; concept of absolute value, ordering of rational numbers; operations on rational numbers; properties of the rationals; solution of first degree equations involving rational coefficients; graphing of temperature and uniform speed.

    3.   Equations with integral coefficients   46 hrs.
          Monominals and polynominals; operations involving polynominals; simplification of polynominals; $(a + b)^2$, $(a + b)(a - b)$, $(a \pm b)^3$, $(a + b)(a^2 \pm ab + b^2)$.

    4.   Factoring of polynominal equations   22 hrs.
          Two methods of finding common factors.

    5.   Rational equations   20 hrs.
          Properties of rational equations; simplification of equations; methods of finding common denominator for several equations; operations on rational expressions.

Eighth Grade

    6. Review   4 hrs.

Geometry   2 hrs/wk   total 68 hrs.

    1. General introduction   16 hrs.
Review of plane figures and solids; concepts of point, line, plane, line segment; comparison of line segments; addition and subtraction of line segments; review circle knowledge; arcs, comparison of arcs; addition and subtraction of arcs; angles; comparison of angles; operations on angles; angles in a plane and space, measurement and classification of angles; geometry as an axiomatic system; discussion of methods of proof in geometry.

    2. Triangles   42 hrs.
Angle defined as a "broken line"; classification of polygons; classification of triangles; symmetry; properties of isosceles triangles; conditions for congruence; properties of exterior angles; triangular inequality, properties of right triangles; locus problems; construction problems - duplication of a given angle, bisect given angles and lines, perpendicular from a given point to a given line.

    3. Survey practice   6 hrs.
Location of intersection point of two lines; use of level; measure and layout of angles; measuring distance between two points with an obstacle between them; measuring the height of a given object when the top is beyond reach.

    4. Review   4 hrs.

Ninth Grade

    Algebra   3 hrs/wk   total 102 hrs.

    1. Rational expressions   22 hrs.
Properties; simplification of expressions; finding common denominator for two or more expressions; operations on rational expressions.

Ninth Grade

2. Ratio   6 hrs.
   Properties of ratio; change of terms in a given
   ratio; properties of the function $y = ax$.

3. First degree equations (1 variable)   26 hrs.
   Concepts of identity and function; properties of
   functions; solution of first degree equations
   with integral coefficients; use of functions in
   problem situations; concept of inequality; first
   degree inequalities.

4. First degree equations (2 variables)   22 hrs.
   Discussion of solution possibilities; solution
   of first degree equations in two or three variables; work with numerical and variable coefficients; graphing of functions including
   $y = ax + b$ and $ax + by = 0$.

5. Square roots   12 hrs.
   Finding roots of perfect squares; approximate
   roots of rational numbers; consideration of the
   equation $ax^2 = c$.

Geometry   3 hrs/wk   total   102 hrs.

1. Parallel lines
   Definitions and properties; methods of determine
   parallelism; construction of a line parallel to
   another through a given point; consideration of
   a perpendicular to a set of parallel lines; relation between the three interior angles of a
   triangle; properties of the "30-60-90" right
   triangle; sum of the interior and exterior angles
   of a polygon.

2. Quadrilaterals   24 hrs.
   Parallelograms; definitions; properties of
   diagonal; center of symmetry; construction of
   parallelogram from given data; properties of
   "diamond" shaped quadrilaterals; division of
   line into proportional parts; properties established by the bisectors of the sides of a triangle and trapezoid.

3. Circle   30 hrs.
   Basic properties; uniqueness of circle through
   three given points; diameter perpendicular to
   chord bisects sector formed; parallel lines

Ninth Grade

intersect equal arcs; division of arc into
equal parts; location of center of given circle;
diameter defined as maximum chord; definition of
tangent; properties of tangents; draw tangent to
a given circle parallel to a given line; relationship between two circles; construction
problems; tangents, sector with given central
angle; locus problems.

4. Inscribe and circumscribe circles in and about
   triangles and quadrilaterals   10 hrs.
   Definitions; inscribe and circumscribe quadrilaterals and polygons; circle must pass through
   vertices of triangle; uniqueness of inscribed
   circle tangent to three sides of triangle; students generalize principles to other polygons;
   location of orthocenter and center of gravity
   for a triangle.

5. Survey Practice   8 hrs.
   Layout of parallel lines; measurement of distance
   between two points with an obstacle between; use
   of compass; determining azimuths; area of polygonal fields.

6. Review   14 hrs.

Tenth Grade

Algebra   4 hr/wk (1st semester), 3 hr/wk (2nd semes-
                                                ter)
                                   total 120 hrs.

1. Powers and roots   36 hrs.
   Positive integral powers; negative powers;
   powers of rational numbers; multiplication and
   division of expressions with the same base;
   find a given root of any number; proof: $\sqrt{2}$ is
   not rational; concept of irrational numbers;
   irrational numbers as infinite nonrepeating
   decimals; use of coordinates to show real numbers; operations involving roots; use of root
   reducing formulae.

2. Factoring of second degree equations   46 hrs.
   Consideration of the equations: $ax^2 + c = 0$,
   $ax^2 + bx = 0$, $x^2 + px + q = 0$, $ax^2 + bx + c = 0$;

Tenth Grade

use of quadratic equations in solving problems;
use of discriminate to describe roots.

3. Functions and graphs   17 hrs.
   Dependent and independent variables; review of
   coordinates; consideration of the graphs of the
   functions: $y = kx$, $y = k/x$, $y = kx + b$,
   $y = ax^2 + bx + c$.

4. Solutions for systems of equations in two variables   18 hrs.
   Methods of solution for the following systems:
   $x \pm y = a$ and $xy = b$; $x^2 \pm y^2 = a$ and $x \pm y = b$.

5. Review   3 hrs.

Geometry   2 hrs/wk (1st semester)
            3 hrs/wk (2nd semester)   total   84 hrs.

1. Similarity   33 hrs.
   Point of projection method to divide given line
   segment into proportional parts; construction of
   line segments in given ratios; principle of
   similarity for triangles; similarity for polygons; similar polygon divided into similar
   triangles; methods of reducing and enlarging
   figures; applied problems involving principles
   of similarity.

2. Trigonometric functions of acute angles 10 hrs.
   Definitions of sine, cosine, tangent and co-
   tangent functions; use of tables; applied problems.

3. Properties of triangles and circles   10 hrs.
   Relationship between the altitude and base of a
   triangle; projections on the hypotenuse; sum of
   the squares of the diagonals of a rhombus; given
   three sides of a triangle find the altitude;
   properties of intersecting chords within a
   circle; construction of line segment of length
   x such that

   $x = \dfrac{ab}{c}$, $x = \sqrt{ab}$, $x = \sqrt{a^2 + b^2}$ where a, b, c,
   are given.

Tenth Grade

4. Area of polygons   12 hrs.
    Concept of area; areas of rectangles, triangles,
    diamonds and trapezoids; construction of a
    square and triangle with area equal to that of a
    given polygon; find area of triangle when lengths
    of three sides given; ratio of areas of similar
    polygons.

5. Survey practice   6 hrs.
    Planemeter practice; make drawings of given
    polygonal plots and calculate area; measure
    distance between two points with obstacles
    between.

6. Review   3 hrs.

Eleventh Grade

Algebra   2 hrs/wk   total 68 hrs.

1. Series   18 hrs.
    Concept of series; examples of series; the nth
    term of a series; arithmetic and geometric
    series; sum of a series to the nth term; concept
    of limit; conversion of repeating decimal to a
    rational number.

2. Advanced theory of powers   8 hrs.
    Zero, negative and fractional powers; irrational
    power.

3. Power and logarithmic functions   40 hrs.
    Graphs and properties of power functions; defini-
    tion of logarithm; graphs of logarithmic func-
    tions; operation on logs; use of log tables.

4. Review   2 hrs.

Geometry   2 hr/wk   total 68 hrs.

Plane Geometry

1. Regular polygons   12 hrs.
    Definition of regular polygons; principles for
    constructing circumscribed and inscribed polygon
    about and within circles; ratio of circumferences
    of two regular polygons with equal sides; given

Eleventh Grade

a circle, find the length of the side of an
inscribed square, hexagon and equilateral triangle; given a circle, construct an inscribed
decagon and calculate the length of its sides;
construct an inscribed regular polygon and
double the number of its sides successively.

2. Circumference and area of a circle   11 hrs.
    Principle of polygon with all interior angles
    less than 180°; definition of circumference;
    ratio of circumference to radii; length of arc
    with central angle · n degrees; area of circle;
    ratio of area of two circles.

Solid Geometry

1. Straight line and plane   43 hrs.
    Properties of a plane; straight lines on differ-
    ent planes; line parallel to a plane; definition
    of parallel planes; properties of plane inter-
    secting two parallel planes; perpendicular and
    inclined lines to a plane; length of perpendicu-
    lar and inclined lines drawn from a point out-
    side the plane; principle of three perpendicular
    lines and its converse; angle between a line and
    a plane; construction problems: (1) given two
    lines in different planes draw a plane through
    one line parallel to the other, (2) through a
    given point construct a plane parallel to a
    given plane, (3) construct a plane perpendicular
    to a line, (4) from a given point construct a
    line perpendicular to a plane; angle between two
    planes; definition of a plane perpendicular to
    two other planes; polyhedron; definition of
    convex multi-plane angle and face angles; proper-
    ties of three face angles; properties of multi-
    plane angles.

2. Review   2 hrs.

Trigonometry   2 hrs/wk   total 68 hrs.

1. Values of trigonometric functions for angles
    between 0 and 360 degrees   18 hrs.
    Definitions of trigonometric functions; find
    angle with known trigonometric function; use of
    tables; relation between trigonometric functions

Eleventh Grade

of complimentary angles; values of trigonometric functions for angles of 30°, 45°, and 60°.

2. Radian measure   3 hrs.
   Concept of radian measure; relationship between radians and degrees; relationship of central angle, radii and intercepted arc.

3. Trigonometric functions of any angle   10 hrs.
   Concept of positive and negative angles; development of formulae to change any angle to form which trigonometric functions apply; periodicity of trigonometric functions; graphs of trigonometric functions.

4. Trigonometric formulae   20 hrs.
   Trigonometric functions for the sum and difference of two angles, double an angle and half an angle; applications.

5. Logarithms of trigonometric functions   36 hrs.
   Use of tables to solve problems.

6. Rectangular pyramids   8 hrs.
   Solution of problems involving rectangular pyramids.

7. Survey practice   4 hrs.

8. Review   8 hrs.

Twelfth Grade

Algebra   2 hrs/wk   total 68 hrs.

1. Combinations, permutations and the Binominal Theorem   12 hrs.
   Possible arrangements of objects; development of the formulae $C_m^n = C_m^{m-n}$; binominal expansions involving positive integer exponents; Newton's equation; find the coefficients of two symmetric terms in a binominal expression; sum of the coefficients of a binominal expansion

2. Complex numbers   10 hrs.
   Complex number; conjugate of a complex number; defining the complex number "0"; equality for

Twelfth Grade

complex numbers; operations on complex numbers; graphic representation of complex numbers; trigonometric expressions for complex numbers.

3. Inequalities   22 hrs.
   Properties of algebraic inequalities; operations on inequalities; solution of first inequalities.

4. Higher degree equations   12 hrs.
   Necessary and sufficient conditions for a multi-term equation to be divided by x-a; solutions of 3rd, 4th, and 6th degree binominal expansions.

5. Review   12 hrs.

Geometry   2 hrs/wk   total 68 hrs.

1. Straight line and plane   6 hrs.
   Multi-face angle; definition of a convex faced angle.

2. Polyhedra   30 hrs.
   Properties of common prisms and pyramids; surface area; concept of volume; consideration of frustra of pyramids and cones; solution of problems by means of trigonometry and geometry.

3. Solids of revolution   20 hrs.
   Properties of cylinders and cones; surface area; volume; frustra; construction layouts for cylinders and cones; definition of a sphere; properties of intersecting and tangent planes to a sphere; surface area of a sphere; concept of a great circle; ratio of the surface area of two spheres; solution of problems by methods of trigonometry and geometry.

4. Review   12 hrs.

Trigonometry   2 hrs/wk   total 68 hrs.

1. General trigonometry   22 hrs.
   Relation between the sides and angles of a tri-angle; review of trigonometric functions; use of trigonometric tables to solve problems.

Twelfth Grade

2. Inverse trigonometric functions   10 hrs.
    Definition and symbols for inverse trigonometric
    functions; the general solution for an inverse
    function; the principle solution.

3. Trigonometric equations   16 hrs.
    General methods to solve trigonometric equa-
    tions.

4. Survey practice   4 hrs.
    Measure distance between two points with
    obstacle between; measure height of object with
    top not accessible; topographic survey.

5. Review   16 hrs.

APPENDIX D

SAMPLE MATHEMATICS EXAMINATIONS

Entrance Examination in Mathematics
Fu Tan University, September 1, 1917*

Geometry

1. Define: axiom, locus, postulate, perpendicular bisector and tangent.

2. State the cases in which two triangles are congruent.

3. Prove: The diagonals of a parallelogram bisect each other.

4. Prove: The perpendicular bisectors of the sides of a triangle are concurrent in a point equidistant from the vertices.

5. Prove: The lines joining the mid-points of the consecutive sides of any quadrilateral form a parallelogram.

6. Prove: The tangents to a circle drawn from an external point are equal, and make equal angles with the line joining the point to the center.

7. The diameter AB and the chord DC (of a circle) are prolonged until they meet at E. Prove EA > EC and EB < ED.

Solid Geometry

1. Prove that a spherical angle is measured by the arc of the great circle described from its vertex as a pole and included between its sides (produce if necessary).

2. The sides of a spherical triangle are 80°, 74° and 128°. The radius of the sphere is 14 ft. Find the area of the polar triangle.

3. Prove that the sum of the sides of a spherical polygon is less than 360°.

Solid Geometry

4. A plane divides the surface of a sphere of radius
    R into two zones such that the surface of the
    greater is the mean proportional between the entire
    surface and the surface of the smaller. Find the
    distance of the plane from the center of the sphere.

5. Find the area of the surface and the volume of a
    sphere if the diameter is 3 ft. 6 inches.

6. Prove that two mutually equilateral triangles on
    the same sphere or equal spheres are mutually equi-
    angular and are equal or symmetrical.

Trigonometry

1. Prove that:

    $(1 + \tan x)(\tan 2x) = \sec 2x$

2. If $\sin x = -5/13$ and x is in the fourth quadrant,
    find the other functions of x.

3. Show that $\tan 3x = \dfrac{3 \tan x - \tan^3 x}{1 - 3 \tan^2 x}$

4. If $Y = \sin^{-1}(1/3)$ find tan Y.

5. Compute the value of $\sin 90° - b \cos 360° + (a - b) \cos 180°$

6. Find the value of x by logarithms if

    $$x = \sqrt[4]{\frac{25243 \times 0.0534}{0.0063 \times 275200}}$$

* Source: Fuh Tan University Catalogue and Directory
           1917-18, p. 49.

1957 College Entrance Mathematics Examination*

1. (a) Simplify $(2\ 7/9)^{\frac{1}{2}} + 0.1^{-2} + 2\ (10/27)^{-2/3}$.

   (b) Find the domain of the inequality: $x^2 + x > 2$.

   (c) Prove: $\cot 22°\ 30' = 1 + 2$.

   (d) Given a polygon ABCD where: AC = BD.
   Let P, O, R, S be the respective median points
   of AB, BC, CD, and DA.
   Prove: PQRS is a diamond shaped polygon.

2. Solve for x and y simultaneously

   $$\log (2x + 1) + \log (y - 2) = 1$$
   $$10^{xy} = 10^x \cdot 10^y$$

3. Given a circle of radii r inscribed in a triangle
   ABC with altitude AD,

   Prove: $AD = \dfrac{2r\ \cos B/2\ \cos C/2}{\sin A/2}$

4. Given an acute triangle ABC, construct a circle using
   side BC of the triangle as a diameter; draw a tangent
   to the circle from vertex A; let the point of
   tangency be D; construct a line segment AE equal in
   length to AD; from E draw a perpendicular to AB
   intersecting the extension of AC at F.
   Prove: (a) AE:AB = AC:AF
          (b) Area of triangle ABC = Area of triangle
            AEF.

5. (a) Show that one of the roots of
   $$x^3 - (\sqrt{2} + 1)\ x^2 + (\sqrt{2} - q)\ x + q = 0 \text{ is } 1.$$

   (b) Suppose the roots of the equation are: Sin A,
   Sin B and Sin C of some triangle ABC. Find the
   value of angles A, B, and C in degrees and the
   value of q.

*Statistics available on this college entrance
examination reveal that one hundred forty six or 49 per
cent of the candidates achieved scores exceeding 50 per
cent. These figures compare with 38 per cent reaching the
same standard on the previous year's examination and was
taken as in indication of improved middle school mathematics teaching. "Peking City 1957 College Entrance Results

in Mathematics," <u>Shuxue Tongbao</u> (May, 1958), pp. 12-13.
For a detailed outline of the examination for matricula-
tion to institutions of higher education in the People's
Republic of China for 1959, see <u>Current Background</u> (Hong
Kong: United States Consulate General, August 17, 1969),
pp. 6-12.

Peking Mathematical Olympiad: 1963*

Junior Level Examination: First Round

1. Ten people are grouped into two clubs, each club con-
    sisting of five members. In each club a president
    and a vice-president are chosen. How many ways can
    this be done?

2. Given: sin a + sin b = p, cos a + cos b = q, find
    the values of sin (a + b) and cox (a + b).

3. Solve the simultaneous equations:

    $$\sqrt{x - 1} \; + \; \sqrt{y - 3} \; = \; \sqrt{x + y}$$
    $$\lg (x - 10) \; + \; \lg (y - 6) \; = \; 1$$

4. The lengths of the sides of a right triangle form
    three consecutive terms of an arithmetic progression.
    Prove that the lengths are in the ratio 3:4:5.

5. Let D be a point on the arc BC of the circumscribed
    circle about the equilateral triangle ABC. Let E be
    the intersection of the lines AB and CD, F the
    intersection of the lines AC and BD.

    Prove $\overline{BC}$ is the geometric mean of $\overline{BE}$ and $\overline{CF}$.
    $[\overline{BC}^2 = \overline{BE} \cdot \overline{CF}]$

Junior Level Examination: Second Round

1. Let $x^3 + bx^2 + cx + d$ be a polynomial with integral
    coefficients, and let bd + cd be odd. Prove the
    polynomial is not the product of two polynomials,
    each with integral coefficients.

2. Suppose five points are given in the plane, no 3 on
    a line, no 4 on a circle. Prove there exists a
    circle through 3 of the points such that of the
    remaining 2 points, one is in the interior and the
    other is in the exterior of the circle.

3. Let P be a point in the interior of a regular hexagon whose sides have length 1. The line segments from P to two vertices have length 13/12 and 5/12 respectively. Determine the lengths of the segments from P to the 4 remaining vertices.

4. Let a be a positive integer, and let $r = \sqrt{a + 1} + \sqrt{a}$. Prove that for any positive integer n there exists a positive integer a_n satisfying:

$$r^{2n} + r^{-2n} = 4\,a_n + 2$$

$$r^n = \sqrt{a_n + 1} + \sqrt{a_n}$$

Senior Level Examination: First Round

1. If $2 \lg(x - 2y) = \lg x + \lg y$, find $x{:}y$.

2. Let r and R be the radii respectively of the inscribed and the circumscribed circles to a regular n-gon whose sides have length a.

   Prove: $r + R = \dfrac{a}{2} \cot \dfrac{\pi}{2n}$

3. Find the coefficient of x^2 in

   $$(1 + x)^3 + (1 + x)^4 + (1 + x)^5 + \cdots + (1 + x)^{n+2}.$$

4. Given a convex n-gon, call the line segment joining two non-adjacent vertices a diagonal. Assume no 3 diagonals intersect in a common point. Find the number of intersections of diagonals (in the interior of the n-gon).

5. A trapezoid is given with parallel edges of lengths a and 2a. A side of the trapezoid has length b and forms an acute angle a with the edge of length 2a. Find the volume of the solid of revolution determined by rotating the trapezoid about the side of length b.

Senior Level Examination: Second Round

1. Let $P(x) = P(x) = A_k X^k + A_{k-1} X^{k-1} + \cdots + A_1 X + A_0$
    be a polynomial with integral coefficients. Suppose
    x_1, x_2, x_3, x_4 are distinct integers such that
    $P(x_1) = 2$ for $i = 1, 2, 3, 4$. Prove that $P(x)$ is
    not 1, 3, 5, 7, or 9 for any integer x.

2. Let 9 points be given in the interior of the unit
    square. Prove there exists a triangle of area $\geq \frac{1}{8}$
    whose vertices are 3 of the 9 points.

3. 2n + 3 points are given in the plane, no 3 on a line,
    no 4 on a circle. Is it possible to find a circle
    through 3 of the points such that of the remaining
    2n points, half are in the interior and half are in
    the exterior of the circle? Prove your answer.

4. 2^n counters are divided into several piles. The
    following defines a move: choose two piles A and B,
    say with p and q counters respectively, $p \geq q$; move
    q counters from A and put them in pile B. Prove
    there exists a finite number of moves such that all
    counters end up in one pile.

* Source: "Results of the Mathematics Contest for Middle
            School Students in Peking Municipality,"
            Shuxue Tongbao (April, 1963), pp. 50-51; and

            "1963 Peking Municipality Mathematical Compe-
            titions," Shuxue Tongbao (May, 1963), back
            cover; other published examination questions
            can be found in: "1957 Peking Municipality
            Mathematical Competitions," Shuxue Tongbao
            (May, 1957), pp. 38-44; "1957 Tiensin, Wuhan
            and Nanking Mathematical Competitions," Shuxue
            Tongbao, (August, 1957), pp. 45-46; and Com-
            pilation of Problems from 1956-57 Mathematical
            Competitions for Middle School Students in
            Shanghai Municipality, op. cit.

APPENDIX E

SAMPLE LESSON PLANS USED IN CANTON MIDDLE SCHOOL
1962 *

Algebra Lesson

Topic: Discriminant of quadratic.

Purpose and Request: 1. Let the students understand what is discriminant of quadratic.

2. Let the students through Discriminant how to decide what kind of number the roots belong.

3. Let the students get the good foundation ready for the discussion of Algebraic equations.

Process:
Organization 1. Let the students be in a good order and be attentive to the teacher.

2. Review and question.

1. What is a quadratic and a quadratic equation?
2. How many roots are there in a quadratic equation?
3. Write a general form of quadratic equation ($ax^2 + bx + c = 0$)
4. If A and B represent the two roots of the equation

then A = ?
     B = ?

Let the student derive:

$$A = \frac{-b + \sqrt{b^2 - 4ac}}{2a} \qquad B = \frac{-b - \sqrt{b^2 - 4ac}}{2a}$$

3. Start the new lesson:
What kind of number is A & B, real or imaginary?
It depends on

1. When $b^2 - 4ac > 0$
    then $b^2 - 4ac$ = ?   (what kind of number)
    $\sqrt{b^2 - 4ac}$ = ?

    and A = $\dfrac{-b + \sqrt{b^2 - 4ac}}{2a}$ = ?   B = $\dfrac{-b - \sqrt{b^2 - 4ac}}{2a}$ = ?

    (two roots are unequal real number)

2. When $b^2 - 4ac = 0$
    then: $b^2 - 4ac$ = ?
    $\sqrt{b^2 - 4ac}$ = ?   (what kind of number)

    and A = $\dfrac{-b + \sqrt{b^2 - 4ac}}{2a}$ = ?   B = $\dfrac{-b - \sqrt{b^2 - 4ac}}{2a}$ = ?

    (two roots are equal real number)

3. When $b^2 - 4ac < 0$
    then: $b^2 - 4ac$ = ?   (what kind of number)
    $\sqrt{b^2 - 4ac}$ = ?

    and A = $\dfrac{-b + \sqrt{b^2 - 4ac}}{2a}$ = ?   B = $\dfrac{-b - \sqrt{b^2 - 4ac}}{2a}$ = ?

    (two roots are conjugate imaginary)

We have to know more about:

1. When $b^2 - 4ac = 0$, then $ax^2 + bx + c$ is a complete square.

2. When a is positive and c is negative, then the two roots must be real numbers because $b^2 - 4ac$ is positive.

3. If a, b, c, are rational numbers and $b^2 - 4ac$ is a complete square, then the roots are rational.

4. We may be sure from the graphs: $y = ax^2 + bx + c$ (review the graphs of quadratic formula)

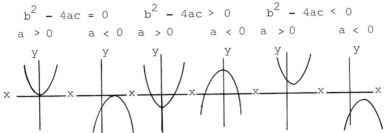

$b^2 - 4ac = 0$ $b^2 - 4ac > 0$ $b^2 - 4ac < 0$

$a > 0$ $a < 0$ $a > 0$ $a < 0$ $a > 0$ $a < 0$

The curve inter-   The curve inter- The curve does not
sects x-axis at   sects x-axis at   intersect x-axis
one point only.   two points.   at any point.

5. Conclusion and review.
    Now we all know what kind of root of a quadratic
    equation depends on the value of $b^2 - 4ac$.

    We call $b^2 - 4ac$ "Discriminant" of quadratic. By
    discriminant we know the properties of the roots of
    a quadratic equation.

    Questions: 1. When $b^2 - 4ac > 0$ what are the roots?
                2. When $b^2 - 4ac = 0$ what are the roots?
                3. When $b^2 - 4ac < 0$ what are the roots?

6. Homework assignment
    1. Prove $x^2 + x + 1 = 0$ which two roots are
                                       imaginary?
    2. $2x^2 + 18 = Kx$ when the roots are equal
                                      what is the value of K?
    3. $3x^2 - 2x + M = 0$ what is the value of M,
                                      when two roots are real,
                                      and imaginary?
    4. Factor: $y^2 + 4xy + 3x^2 - x - 3y - 4$

Geometry Lesson

TOPIC: The sum of the angles of a triangle is equal to a
        straight angle.

PURPOSE AND REQUEST: To make the students assure that
        the sum of the angles of a triangle is equal to a
        straight angle only.
PROCESS:
Organization, 1. Let the students be in a good order and
                  be attentive to the teacher.

2. Review and question:

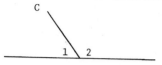

AB is a straight line
$\angle 1 + \angle 2 = ?°$

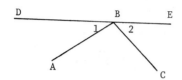

DE is a straight line
$\angle 1 + \angle ABC + \angle 2 = ?°$

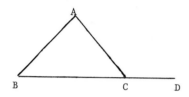

Extend BC to D
What is the relation
of
$\angle A, \angle B$ and $\angle ACD$.

3. Start the lesson
    Write the topic and the triangle ABC on the blackboard.
    Given: $\triangle ABC$
    To prove: $\angle A + \angle B + \angle C = 1$ straight angle.

    Proof 1:   (Introduce the student to think out to
                construct a straight line through the
                vertex B and // AC)

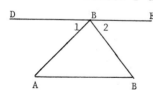

What is the relation of $\angle 1$ and $\angle A$
Why?
What is the relation of $\angle 2$ and $\angle C$
Why?

$\angle 1 + \angle ABC + \angle 2 = ?°$ Why
$\angle A + \angle B + \angle C = ?°$ Why
    (Refer what they have just
    reviewed)

Therefore $\angle A + \angle B + \angle C = 1$ straight angle.

Proof 2

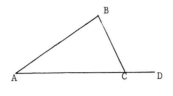

Draw the triangle ABC on the blackboard. If AC is extended to D
Then:

1. What is the relation of ∠A, ∠B, and ∠BCD?
    Ans: ∠A + ∠B = ∠BCD
    (Refer what they have just reviewed)
2. ∠BCD + ∠ACB = ?   Why?
    (Refer what they have just reviewed)
3. ∠A + ∠B + ∠ACB = ?   Why?

Therefore ∠A + ∠B + ∠C = 1 straight angle.

4. Conclusion and Review:
    Either above two proofs we know:
    In any triangle the sum of the three angles is equal to a straight angle or 180°.

Derive the following corollaries by questioning:

1. The sum of any two angles of a triangle is less than a straight angle.
2. There is only one right angle or obtuse angle in a triangle.
3. The two acute angles of a right triangle are complementary.
4. Each angle of a equiangular triangle is 60°.
5. If two angles of a triangle are respectively equal to two angles of another triangle, the third angles are equal.
6. If two angles of a triangle and a side opposite one of them are respectively equal to the two angles of another triangle and a corresponding side, the two triangles are congruent.
7. If an acute angle and hypotenuse of a right triangle are respectively equal to the acute angle and the hypotenuse of another right triangle, the two are congruent.
8. If an acute angle and a side of a right triangle are respectively equal to an acute angle and a side of another right triangle, then the two triangles are congruent.
9. There is only one perpendicular that can be drawn to a given line through a given point outside the line.

5. Homework assignment.

    1. Given a vertex angle of an isosceles triangle, construct a base triangle.
    2. In triangle ABC, the bisector of angle A meets BC at E and extend BC to D. To prove
        $$\angle AED = \angle ABD + \angle ACD$$
    3. The sum of two exterior angles that is found by both extension of hypotenuse and two adjacent sides is three right angles.
    4. In a right triangle, $\angle A$ is a right angle, BD is a bisector of $\angle B$, AF $\perp$ BC, AF intersects BD at E, to prove AD = AE.
    5. In triangle ABC from B draw BE perpendicular to the bisector of $\angle A$ at E, then $\angle ABE$ is equal to the half of the sum of $\angle B$ and $\angle C$ and $\angle EBC$ is equal to the half of the difference of $\angle B$ and $\angle C$.

        *Contributed by former Canton middle school teacher.

INDEX

Japanese, 97, 119, 289;
  influence on education,
  54, 69; middle school
  mathematics, 65; teachers,
  60; textbooks, 60
Jesuits, 17. See also
  Ricci, Matteo
Junior normal school, 208

K'ang-hsü, 41
Kamenosuke, Nagesawa, 259
Key schools, 181-84, 186,
  198, 206, 214, 298;
  universities, 214
Kiangnan translation
  bureau, 48
Kindergarten, 55
Kirin Normal University,
  175-76
Kuang Han, 165
Kuei-liang, 39
Kuomintang (Nationalists),
  94-103, 295-99; middle
  school mathematics, 99-
  102

Laboratory teaching, 88-89
League of Nations, 96, 296
Lesson plans, 138, 351-56
Li Hung-chang, 40; advocate
  of mathematics studies,
  46
Li Kung, 26
Li Shan-lan, 36-39, 43, 47;
  translator of Euclid, 36-
  39
Lin Ch'eng-chu, 177
Lin Feng, 178
Lin Piao, 192
Linear programming, 177
Lishu Plan, 196-99
Literacy, 123; classes, 160
Liu Ching-ho, 228, 237
Liu Shih, 130
Logarithms, 17, 67
Long March, 107
Lo-shu, 11. See also Magic
  squares
Lu Chung-heng, 232

Lu Ting-yi, 170, 171, 181;
  comments on productive
  labor, 157; Speech at
  Second National People's
  Congress, 3

Magic squares, 12
Manchu government, 1, 58;
  education, 53-68; middle
  school curriculum, 57;
  school mathematics, 56.
  See also Ch'ing dynasty
Mao thought, 190, 194,
  245, 268; incorporated
  into texts, 216; in
  Lishu Plan, 197
Mao Tse-tung, 106, 107,
  185, 190, 194, 195, 206,
  290; epistemological
  philosophy, 191-192, educational philosophy, 107-
  110, 126; educational
  reform, 137, 153, 183;
  On Practice, 107; "Sixty
  Articles of Working
  Methods" directive, 156
Mao Yu-yen, 236
Marxist-Leninism, 155, 301;
  required study for teachers, 211
Mateer, Charles, Rev., 49-
  50
Mathematical Olympiad, 7,
  153, 246-257, 298;
  Peking (1963), 348-350;
  Results of, 250, 255-57;
  student study groups, 251
Mathematiches Koye Prosveschenia, 144
Mathematics, 20, 24, 36,
  53, 47, 307; at Tsinchow
  Textile and Machine
  Manufacturing School,
  161-62; enrichment mate-
  rials, 268-69; in border
  regions, 120-22; journals,
  144; Kuomintang Program,
  99-101; Nanking Teacher
  College revision, 74-75;

Surveying activities, 91,
140, 200, 211, 240, 283-
87; extracurricular, 284;
in mathematics curriculum,
303
Surveys of China Mainland
Press, 9

Tai-mao, 11
T'ai-p'ing Rebellion, 34,
36, 37
T'ai-tsu, 25
Taoism, 22; knowledge by
intuition, 33
Tawney, R. H., 96
Teacher "research" teams,
138, 195, 199
Teacher shortage, 7, 163,
222-23
Teacher training, 134, 144-
47, 208-224, 304-5;
institutions, 331-34;
mathematics curricula,
212-13, 215-19; restruc-
turing under Soviet influ-
ence, 209; revisions of
1956, 148. See also
Spare-time education
Teachers College, Columbia
University, 72
Teaching conditions, 144,
189, 193; during Cultural
Revolution, 194-96; in
traditional schools, 27;
in Yenan, 111. See also
Formalism; Memorization
Teaching methods, 87, 89;
Lishu Plan, 197; under
traditional system, 27
Technical education, 2, 137
Television universities,
219-221, 305
"Ten Great Years", 223
Ten Thousand Word Letter, 24
Terman, E. L., 86
Textbooks, 87, 258, 300;
American, 329-330; criticized
for theoretical outlook, 58;
in border regions, 118-22,
124; indoctrination in, 203,
275-76; lack of in Yenan,

111; re-editing during
Cultural Revolution, 198;
revision of, 138, 140,
153, 173; self-study,
261-62. See also Foreign
influence
"Three-in-one" alliances,
195
Three Principles of the
People, 94, 95, 102
Tientsin Technical Middle
School, 187-8
Topology, 253
Translation, 40; Kiangnan
Translation Bureau, 48;
of Soviet texts, 137,
138-9, 150; of western
texts, 47-48. See also
Missionaries
Trigonometry, 44, 57, 79,
166, 200, 283, 341;
extracurricular work in,
239; Jesuit influence,
17; on mathematical
olympiad, 247; reforms of
1922, 80-81
Trimetrical Classic, 28
Tseng Chao-lun, 135
Tseng K'en-cheng, 252, 253
Tseng Kuo-fan, 35, 41;
educational reformer,
36; sponsor of Euclid's
Elements, 36
Ts'e-yüan Hai-ching (Sea
Mirror of Circle Measure-
ments), 16
Tsinchow Textile and Machine
Manufacturing School, 160
Tsung-li yamen, 39
Tuan Hsueh-fu, 247; texts
for youth, 252-53
T'ung-wen-kuan, 39-45;
curriculum of, 44
Twiss, George, 87-88
University, 26, 55; Commu-
nist Labor University,
159; consolidation of,
135; language requirements,
182; pedagogical, 207,
208; People's University,
152; "red and expert", 158
Vector algebra, 264